# DESIGN OF THE REACTOR COOLANT SYSTEM AND ASSOCIATED SYSTEMS FOR NUCLEAR POWER PLANTS

The following States are Members of the International Atomic Energy Agency:

AFGHANISTAN
ALBANIA
ALGERIA
ANGOLA
ANTIGUA AND BARBUDA
ARGENTINA
ARMENIA
AUSTRALIA
AUSTRIA
AZERBAIJAN
BAHAMAS
BAHRAIN
BANGLADESH
BARBADOS
BELARUS
BELGIUM
BELIZE
BENIN
BOLIVIA, PLURINATIONAL
   STATE OF
BOSNIA AND HERZEGOVINA
BOTSWANA
BRAZIL
BRUNEI DARUSSALAM
BULGARIA
BURKINA FASO
BURUNDI
CAMBODIA
CAMEROON
CANADA
CENTRAL AFRICAN
   REPUBLIC
CHAD
CHILE
CHINA
COLOMBIA
CONGO
COSTA RICA
CÔTE D'IVOIRE
CROATIA
CUBA
CYPRUS
CZECH REPUBLIC
DEMOCRATIC REPUBLIC
   OF THE CONGO
DENMARK
DJIBOUTI
DOMINICA
DOMINICAN REPUBLIC
ECUADOR
EGYPT
EL SALVADOR
ERITREA
ESTONIA
ESWATINI
ETHIOPIA
FIJI
FINLAND
FRANCE
GABON
GEORGIA

GERMANY
GHANA
GREECE
GRENADA
GUATEMALA
GUYANA
HAITI
HOLY SEE
HONDURAS
HUNGARY
ICELAND
INDIA
INDONESIA
IRAN, ISLAMIC REPUBLIC OF
IRAQ
IRELAND
ISRAEL
ITALY
JAMAICA
JAPAN
JORDAN
KAZAKHSTAN
KENYA
KOREA, REPUBLIC OF
KUWAIT
KYRGYZSTAN
LAO PEOPLE'S DEMOCRATIC
   REPUBLIC
LATVIA
LEBANON
LESOTHO
LIBERIA
LIBYA
LIECHTENSTEIN
LITHUANIA
LUXEMBOURG
MADAGASCAR
MALAWI
MALAYSIA
MALI
MALTA
MARSHALL ISLANDS
MAURITANIA
MAURITIUS
MEXICO
MONACO
MONGOLIA
MONTENEGRO
MOROCCO
MOZAMBIQUE
MYANMAR
NAMIBIA
NEPAL
NETHERLANDS
NEW ZEALAND
NICARAGUA
NIGER
NIGERIA
NORTH MACEDONIA
NORWAY
OMAN

PAKISTAN
PALAU
PANAMA
PAPUA NEW GUINEA
PARAGUAY
PERU
PHILIPPINES
POLAND
PORTUGAL
QATAR
REPUBLIC OF MOLDOVA
ROMANIA
RUSSIAN FEDERATION
RWANDA
SAINT LUCIA
SAINT VINCENT AND
   THE GRENADINES
SAN MARINO
SAUDI ARABIA
SENEGAL
SERBIA
SEYCHELLES
SIERRA LEONE
SINGAPORE
SLOVAKIA
SLOVENIA
SOUTH AFRICA
SPAIN
SRI LANKA
SUDAN
SWEDEN
SWITZERLAND
SYRIAN ARAB REPUBLIC
TAJIKISTAN
THAILAND
TOGO
TRINIDAD AND TOBAGO
TUNISIA
TURKEY
TURKMENISTAN
UGANDA
UKRAINE
UNITED ARAB EMIRATES
UNITED KINGDOM OF
   GREAT BRITAIN AND
   NORTHERN IRELAND
UNITED REPUBLIC
   OF TANZANIA
UNITED STATES OF AMERICA
URUGUAY
UZBEKISTAN
VANUATU
VENEZUELA, BOLIVARIAN
   REPUBLIC OF
VIET NAM
YEMEN
ZAMBIA
ZIMBABWE

The Agency's Statute was approved on 23 October 1956 by the Conference on the Statute of the IAEA held at United Nations Headquarters, New York; it entered into force on 29 July 1957. The Headquarters of the Agency are situated in Vienna. Its principal objective is "to accelerate and enlarge the contribution of atomic energy to peace, health and prosperity throughout the world".

# DESIGN OF THE REACTOR COOLANT SYSTEM AND ASSOCIATED SYSTEMS FOR NUCLEAR POWER PLANTS

## SPECIFIC SAFETY GUIDE

INTERNATIONAL ATOMIC ENERGY AGENCY
VIENNA, 2020

**IAEA Library Cataloguing in Publication Data**

Names: International Atomic Energy Agency.
Title: Design of the reactor coolant system and associated systems for nuclear power plants / International Atomic Energy Agency.
Description: Vienna : International Atomic Energy Agency, 2020. | Series: IAEA safety standards series, ISSN 1020–525X ; no. GSS-56 | Includes bibliographical references.
Identifiers: IAEAL 20-01291 | ISBN 978–92–0–105719–8 (paperback : alk. paper) | ISBN 978–92–0–161019–5 (pdf)
Subjects: LCSH: Nuclear power plants — Design and construction. | Nuclear reactors — Cooling. | Nuclear power plants — Safety measures.
Classification: UDC 621.039.534 | STI/PUB/1878

# FOREWORD

The IAEA's Statute authorizes the Agency to "establish or adopt... standards of safety for protection of health and minimization of danger to life and property" — standards that the IAEA must use in its own operations, and which States can apply by means of their regulatory provisions for nuclear and radiation safety. The IAEA does this in consultation with the competent organs of the United Nations and with the specialized agencies concerned. A comprehensive set of high quality standards under regular review is a key element of a stable and sustainable global safety regime, as is the IAEA's assistance in their application.

The IAEA commenced its safety standards programme in 1958. The emphasis placed on quality, fitness for purpose and continuous improvement has led to the widespread use of the IAEA standards throughout the world. The Safety Standards Series now includes unified Fundamental Safety Principles, which represent an international consensus on what must constitute a high level of protection and safety. With the strong support of the Commission on Safety Standards, the IAEA is working to promote the global acceptance and use of its standards.

Standards are only effective if they are properly applied in practice. The IAEA's safety services encompass design, siting and engineering safety, operational safety, radiation safety, safe transport of radioactive material and safe management of radioactive waste, as well as governmental organization, regulatory matters and safety culture in organizations. These safety services assist Member States in the application of the standards and enable valuable experience and insights to be shared.

Regulating safety is a national responsibility, and many States have decided to adopt the IAEA's standards for use in their national regulations. For parties to the various international safety conventions, IAEA standards provide a consistent, reliable means of ensuring the effective fulfilment of obligations under the conventions. The standards are also applied by regulatory bodies and operators around the world to enhance safety in nuclear power generation and in nuclear applications in medicine, industry, agriculture and research.

Safety is not an end in itself but a prerequisite for the purpose of the protection of people in all States and of the environment — now and in the future. The risks associated with ionizing radiation must be assessed and controlled without unduly limiting the contribution of nuclear energy to equitable and sustainable development. Governments, regulatory bodies and operators everywhere must ensure that nuclear material and radiation sources are used beneficially, safely and ethically. The IAEA safety standards are designed to facilitate this, and I encourage all Member States to make use of them.

# THE IAEA SAFETY STANDARDS

## BACKGROUND

Radioactivity is a natural phenomenon and natural sources of radiation are features of the environment. Radiation and radioactive substances have many beneficial applications, ranging from power generation to uses in medicine, industry and agriculture. The radiation risks to workers and the public and to the environment that may arise from these applications have to be assessed and, if necessary, controlled.

Activities such as the medical uses of radiation, the operation of nuclear installations, the production, transport and use of radioactive material, and the management of radioactive waste must therefore be subject to standards of safety.

Regulating safety is a national responsibility. However, radiation risks may transcend national borders, and international cooperation serves to promote and enhance safety globally by exchanging experience and by improving capabilities to control hazards, to prevent accidents, to respond to emergencies and to mitigate any harmful consequences.

States have an obligation of diligence and duty of care, and are expected to fulfil their national and international undertakings and obligations.

International safety standards provide support for States in meeting their obligations under general principles of international law, such as those relating to environmental protection. International safety standards also promote and assure confidence in safety and facilitate international commerce and trade.

A global nuclear safety regime is in place and is being continuously improved. IAEA safety standards, which support the implementation of binding international instruments and national safety infrastructures, are a cornerstone of this global regime. The IAEA safety standards constitute a useful tool for contracting parties to assess their performance under these international conventions.

## THE IAEA SAFETY STANDARDS

The status of the IAEA safety standards derives from the IAEA's Statute, which authorizes the IAEA to establish or adopt, in consultation and, where appropriate, in collaboration with the competent organs of the United Nations and with the specialized agencies concerned, standards of safety for protection of health and minimization of danger to life and property, and to provide for their application.

With a view to ensuring the protection of people and the environment from harmful effects of ionizing radiation, the IAEA safety standards establish fundamental safety principles, requirements and measures to control the radiation exposure of people and the release of radioactive material to the environment, to restrict the likelihood of events that might lead to a loss of control over a nuclear reactor core, nuclear chain reaction, radioactive source or any other source of radiation, and to mitigate the consequences of such events if they were to occur. The standards apply to facilities and activities that give rise to radiation risks, including nuclear installations, the use of radiation and radioactive sources, the transport of radioactive material and the management of radioactive waste.

Safety measures and security measures[1] have in common the aim of protecting human life and health and the environment. Safety measures and security measures must be designed and implemented in an integrated manner so that security measures do not compromise safety and safety measures do not compromise security.

The IAEA safety standards reflect an international consensus on what constitutes a high level of safety for protecting people and the environment from harmful effects of ionizing radiation. They are issued in the IAEA Safety Standards Series, which has three categories (see Fig. 1).

**Safety Fundamentals**

Safety Fundamentals present the fundamental safety objective and principles of protection and safety, and provide the basis for the safety requirements.

**Safety Requirements**

An integrated and consistent set of Safety Requirements establishes the requirements that must be met to ensure the protection of people and the environment, both now and in the future. The requirements are governed by the objective and principles of the Safety Fundamentals. If the requirements are not met, measures must be taken to reach or restore the required level of safety. The format and style of the requirements facilitate their use for the establishment, in a harmonized manner, of a national regulatory framework. Requirements, including numbered 'overarching' requirements, are expressed as 'shall' statements. Many requirements are not addressed to a specific party, the implication being that the appropriate parties are responsible for fulfilling them.

**Safety Guides**

Safety Guides provide recommendations and guidance on how to comply with the safety requirements, indicating an international consensus that it

---

[1] See also publications issued in the IAEA Nuclear Security Series.

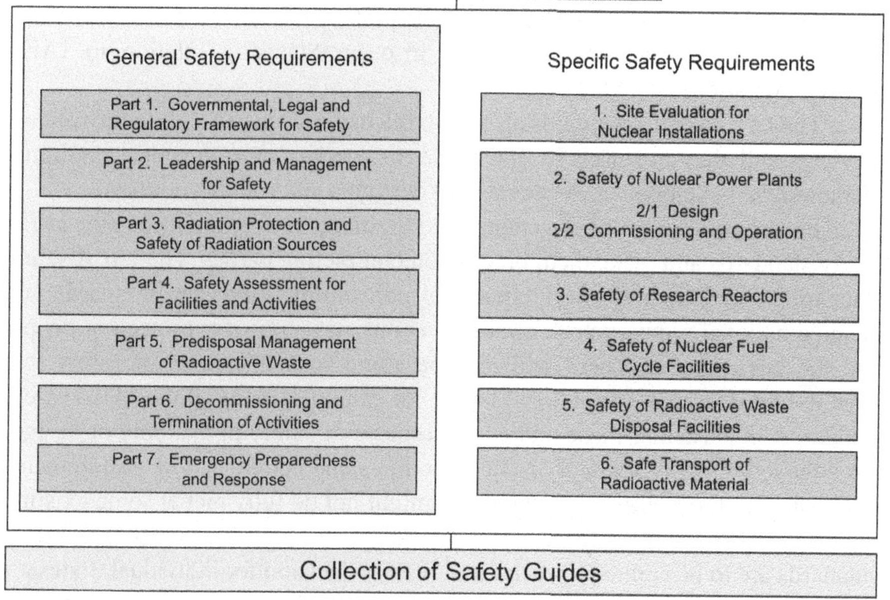



*FIG. 1.   The long term structure of the IAEA Safety Standards Series.*

is necessary to take the measures recommended (or equivalent alternative measures). The Safety Guides present international good practices, and increasingly they reflect best practices, to help users striving to achieve high levels of safety. The recommendations provided in Safety Guides are expressed as 'should' statements.

## APPLICATION OF THE IAEA SAFETY STANDARDS

The principal users of safety standards in IAEA Member States are regulatory bodies and other relevant national authorities. The IAEA safety standards are also used by co-sponsoring organizations and by many organizations that design, construct and operate nuclear facilities, as well as organizations involved in the use of radiation and radioactive sources.

The IAEA safety standards are applicable, as relevant, throughout the entire lifetime of all facilities and activities — existing and new — utilized for peaceful purposes and to protective actions to reduce existing radiation risks. They can be used by States as a reference for their national regulations in respect of facilities and activities.

The IAEA's Statute makes the safety standards binding on the IAEA in relation to its own operations and also on States in relation to IAEA assisted operations.

The IAEA safety standards also form the basis for the IAEA's safety review services, and they are used by the IAEA in support of competence building, including the development of educational curricula and training courses.

International conventions contain requirements similar to those in the IAEA safety standards and make them binding on contracting parties. The IAEA safety standards, supplemented by international conventions, industry standards and detailed national requirements, establish a consistent basis for protecting people and the environment. There will also be some special aspects of safety that need to be assessed at the national level. For example, many of the IAEA safety standards, in particular those addressing aspects of safety in planning or design, are intended to apply primarily to new facilities and activities. The requirements established in the IAEA safety standards might not be fully met at some existing facilities that were built to earlier standards. The way in which IAEA safety standards are to be applied to such facilities is a decision for individual States.

The scientific considerations underlying the IAEA safety standards provide an objective basis for decisions concerning safety; however, decision makers must also make informed judgements and must determine how best to balance the benefits of an action or an activity against the associated radiation risks and any other detrimental impacts to which it gives rise.

## DEVELOPMENT PROCESS FOR THE IAEA SAFETY STANDARDS

The preparation and review of the safety standards involves the IAEA Secretariat and five safety standards committees, for emergency preparedness and response (EPReSC) (as of 2016), nuclear safety (NUSSC), radiation safety (RASSC), the safety of radioactive waste (WASSC) and the safe transport of radioactive material (TRANSSC), and a Commission on Safety Standards (CSS) which oversees the IAEA safety standards programme (see Fig. 2).

All IAEA Member States may nominate experts for the safety standards committees and may provide comments on draft standards. The membership of the Commission on Safety Standards is appointed by the Director General and includes senior governmental officials having responsibility for establishing national standards.

A management system has been established for the processes of planning, developing, reviewing, revising and establishing the IAEA safety standards.

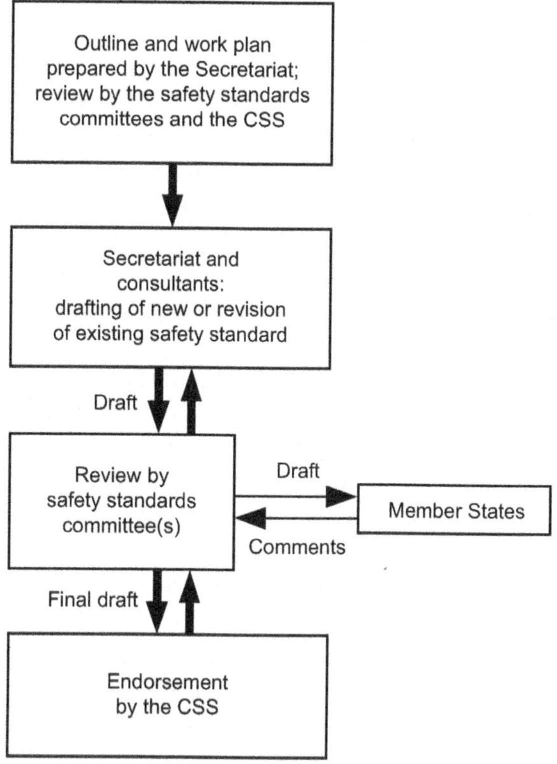


*FIG. 2. The process for developing a new safety standard or revising an existing standard.*

It articulates the mandate of the IAEA, the vision for the future application of the safety standards, policies and strategies, and corresponding functions and responsibilities.

## INTERACTION WITH OTHER INTERNATIONAL ORGANIZATIONS

The findings of the United Nations Scientific Committee on the Effects of Atomic Radiation (UNSCEAR) and the recommendations of international expert bodies, notably the International Commission on Radiological Protection (ICRP), are taken into account in developing the IAEA safety standards. Some safety standards are developed in cooperation with other bodies in the United Nations system or other specialized agencies, including the Food and Agriculture Organization of the United Nations, the United Nations Environment Programme, the International Labour Organization, the OECD Nuclear Energy Agency, the Pan American Health Organization and the World Health Organization.

# INTERPRETATION OF THE TEXT

Safety related terms are to be understood as defined in the IAEA Safety Glossary (see http://www-ns.iaea.org/standards/safety-glossary.htm). Otherwise, words are used with the spellings and meanings assigned to them in the latest edition of The Concise Oxford Dictionary. For Safety Guides, the English version of the text is the authoritative version.

The background and context of each standard in the IAEA Safety Standards Series and its objective, scope and structure are explained in Section 1, Introduction, of each publication.

Material for which there is no appropriate place in the body text (e.g. material that is subsidiary to or separate from the body text, is included in support of statements in the body text, or describes methods of calculation, procedures or limits and conditions) may be presented in appendices or annexes.

An appendix, if included, is considered to form an integral part of the safety standard. Material in an appendix has the same status as the body text, and the IAEA assumes authorship of it. Annexes and footnotes to the main text, if included, are used to provide practical examples or additional information or explanation. Annexes and footnotes are not integral parts of the main text. Annex material published by the IAEA is not necessarily issued under its authorship; material under other authorship may be presented in annexes to the safety standards. Extraneous material presented in annexes is excerpted and adapted as necessary to be generally useful.

# CONTENTS

# 1. INTRODUCTION

## BACKGROUND

1.1. This Safety Guide on the design of the reactor coolant system and associated systems for nuclear power plants provides recommendations on how to meet the requirements of IAEA Safety Standards Series No. SSR-2/1 (Rev. 1), Safety of Nuclear Power Plants: Design [1], in relation to the cooling systems for nuclear power plants.

1.2. This Safety Guide is a revision of IAEA Safety Standards Series No. NS-G-1.9, Design of the Reactor Coolant System and Associated Systems in Nuclear Power Plants[1], which it supersedes.

## OBJECTIVE

1.3. The objective of this Safety Guide is to provide recommendations for the design of the reactor coolant system and associated systems, as described in Section 2, to meet the requirements for these systems established in SSR-2/1 (Rev. 1) [1].

1.4. The recommendations provided in this Safety Guide are aimed at regulatory bodies, nuclear power plant designers and nuclear power plant licensees.

1.5. The terms used in this Safety Guide are to be understood as defined and explained in the IAEA Safety Glossary [2].

## SCOPE

1.6. This Safety Guide applies primarily to land based stationary nuclear power plants with water cooled reactors, designed for electricity generation. It is recognized that for other reactor types, including innovative developments in future systems, some parts of the Safety Guide might not be applicable or might need some judgement to be applied in their interpretation.

---

[1] INTERNATIONAL ATOMIC ENERGY AGENCY, Design of the Reactor Coolant System and Associated Systems in Nuclear Power Plants, IAEA Safety Standards Series No. NS-G-1.9, IAEA, Vienna (2004).

1.7.  The recommendations provided in this Safety Guide are targeted primarily at new nuclear power plants. For nuclear power plants designed in accordance with earlier standards, it is expected that in the safety assessments of such designs a comparison will be made with the current standards (e.g. as part of the periodic safety reassessment for the plant) to determine whether the safe operation of the plant could be further enhanced by means of reasonably practicable safety improvements (see para. 1.3 of SSR-2/1 (Rev. 1) [1]).

1.8.  This Safety Guide covers the reactor coolant system and associated systems, including the ultimate heat sink, as described in Section 4. It covers design considerations for the reactor coolant system and associated systems that are common for the reactor types described in para. 1.6. The scope does not extend to the detailed design of specific components.

1.9.  Section 2 describes the extent of the reactor coolant system and associated systems that are addressed in this Safety Guide. To be independent of individual designs to the extent practicable, design recommendations are given on the basis of the safety functions to be fulfilled by the systems.

1.10. Design limits and engineering criteria, together with the system parameters that should be used to verify these limits and criteria, are specific to individual designs for nuclear power plants and are therefore outside the scope of this Safety Guide. However, qualitative recommendations on these topics are provided.

1.11. Fuel elements and control rods for controlling the core reactivity and shutting down the reactor are not addressed in this Safety Guide.


STRUCTURE

1.12. Section 2 describes the extent of the reactor coolant system and associated systems that are addressed in this Safety Guide. Section 3 provides generic design recommendations for the reactor coolant system and associated systems, which apply to pressurized water reactor (PWR), boiling water reactor (BWR) and pressurized heavy water reactor (PHWR) technologies. Section 4 provides design recommendations for the different heat transfer chains and generic considerations for the ultimate heat sink. Section 5 provides supplementary design recommendations that are specific to reactor coolant systems for PWR, BWR and PHWR technologies. Sections 6–8 provide supplementary design recommendations that are specific to associated systems for PWR, BWR and PHWR technologies, respectively.

# 2. EXTENT OF THE REACTOR COOLANT SYSTEM AND ASSOCIATED SYSTEMS

2.1. The recommendations provided in this Safety Guide apply to the reactor coolant system and associated systems designed to fulfil the following functions:

(a)     To provide confinement of radioactive material for the protection of workers, the public and the environment;
(b)     To provide and maintain adequate core cooling conditions to ensure compliance with fuel design limits in operational states;
(c)     To maintain sufficient coolant inventory and cooling conditions to prevent significant fuel damage in design basis accidents and to mitigate the consequences of design extension conditions to the extent practicable;
(d)     To remove decay heat from the core and to transfer residual heat from the reactor coolant system to the ultimate heat sink in operational states and in accident conditions;
(e)     To prevent an uncontrolled loss of the coolant inventory at the reactor coolant pressure boundary;
(f)     To limit overpressure of the reactor coolant system in operational states, design basis accidents and design extension conditions without significant fuel degradation;
(g)     To shut down the reactor and to control the core reactivity to ensure compliance with fuel design limits in operational states and in accident conditions;
(h)     To perform depressurization of the reactor coolant system in accident conditions.

## REACTOR COOLANT SYSTEM

2.2. For all reactor types, the reactor coolant system includes the components necessary to provide and maintain adequate core cooling conditions (pressure, temperature and coolant flow rate) for the fuel in power operation. As stated in para. 1.11, fuel elements and control rods for controlling the core reactivity and shutting down the reactor are not addressed in this Safety Guide.

2.3. For all water cooled reactor types, the reactor coolant pressure boundary of the reactor coolant system extends up to and includes the outermost isolation device(s).

2.4. For indirect cycle reactors (i.e. PWRs), the pressure retaining boundary of the reactor coolant system includes the primary side of the steam generators (see Section 6). For direct cycle reactors (i.e. BWRs), the pressure retaining boundary of the reactor coolant system also includes the primary coolant recirculation system and the steam lines and feedwater lines up to and including the outermost containment isolation valve (see Section 7). Specific features of PHWRs are described in Section 8.

## SYSTEMS FOR HEAT REMOVAL IN SHUTDOWN CONDITIONS

2.5. These systems are designed to remove residual heat from the reactor coolant system during shutdown. They include systems designed to cool down the reactor coolant system to a cold shutdown condition, including to a refuelling condition for PWRs and BWRs.

## SYSTEMS FOR COOLANT INVENTORY CONTROL IN OPERATIONAL STATES

2.6. These systems are designed to control the reactor coolant inventory and to compensate for leakages in operational states.

## SYSTEMS FOR CONTROL OF CORE REACTIVITY IN OPERATIONAL STATES

2.7. These systems are designed to accommodate slow reactivity changes (including control of the core power distribution) in power operation and to maintain subcriticality in shutdown conditions.

## SYSTEMS FOR CORE COOLING AND RESIDUAL HEAT REMOVAL IN ACCIDENT CONDITIONS

2.8. These systems are as follows:

(a) Systems designed to remove decay heat from the core in accident conditions with or without a loss of the integrity of the reactor coolant system;

(b)     Systems designed to cool the reactor coolant system in accident conditions until safe shutdown conditions are reached and to transfer residual heat from the reactor coolant system to the ultimate heat sink;

(c)     Systems designed to maintain safe shutdown conditions in the long term.


## SYSTEMS FOR CONTROL OF CORE REACTIVITY IN ACCIDENT CONDITIONS

2.9.   These systems are designed to achieve the following:

(a)     To shut down the reactor;
(b)     To stop an uncontrolled or excessive positive reactivity insertion caused by accident conditions;
(c)     To limit fuel damage in the event of an anticipated transient without scram;
(d)     To ensure control of core reactivity until safe shutdown conditions are reached in accident conditions.


## ULTIMATE HEAT SINK AND RESIDUAL HEAT TRANSFER SYSTEMS IN ALL PLANT STATES

2.10. The ultimate heat sink is a medium into which the transferred residual heat can always be accepted, even if all other means of removing the heat have been lost or are insufficient [2]. The ultimate heat sink is usually a body of water (including groundwater) or the atmosphere.

2.11. Residual heat transfer systems include systems designed to transfer heat from the residual heat removal systems to the ultimate heat sink.

2.12. The capabilities to discharge residual heat to the ultimate heat sink are based on at least one heat sink and one heat transfer chain always being available for the different shutdown conditions.

# 3. DESIGN BASIS OF THE REACTOR COOLANT SYSTEM AND ASSOCIATED SYSTEMS

3.1. This section provides generic recommendations for design that are common to the reactor coolant system and associated systems and that are applicable to all water cooled reactors. Design considerations that are specific to a particular reactor technology are described in Section 6 for PWRs, Section 7 for BWRs and Section 8 for PHWRs.

## GENERAL

3.2. A number of reactor coolant systems and associated systems are design dependent and can have different design principles (e.g. the use of active or passive systems for emergency core cooling or for removing residual heat). Nevertheless, systems that fulfil the same safety function in different technologies should be designed in compliance with similar general design requirements.

3.3. The design of the reactor coolant system and associated systems is required to meet Requirements 1–3 of SSR-2/1 (Rev. 1) [1] on the management of safety in design. The design process should also meet the requirements established in IAEA Safety Standards Series No. GSR Part 2, Leadership and Management for Safety [3]. The recommendations provided in IAEA Safety Standards Series Nos GS-G-3.1, Application of the Management System for Facilities and Activities [4], and GS-G-3.5, The Management System for Nuclear Installations [5], should also be taken into account.

3.4. The design of the reactor coolant system and associated systems should be conducted taking into account requirements and recommendations for both safety and nuclear security. Safety measures and security measures should be designed and applied in an integrated manner, and as far as possible in a complementary manner, so that security measures do not compromise safety and safety measures do not compromise security. Recommendations for nuclear security are provided in Ref. [6].

3.5.   The reactor coolant system and associated systems are required to be designed in compliance with Requirements 47–53 of SSR-2/1 (Rev. 1) [1], with account taken of the other requirements of SSR-2/1 (Rev. 1) [1] that are relevant for:

(a)   Protection of workers, the public and the environment in all plant states against harmful effects of ionizing radiation;
(b)   Adequate reliability of the different systems;
(c)   Practical elimination of the possibility of conditions arising that could lead to an early radioactive release or a large radioactive release.

3.6.   To achieve the above mentioned objectives, the reactor coolant system and associated systems should be designed to fulfil the functions listed in para. 2.1 of this Safety Guide.

3.7.   A design basis should be defined for every structure, system and component and should specify the following:

(a)   Function(s) to be performed by the structure, system or component;
(b)   Postulated initiating events that the structure, system or component has to cope with;
(c)   Loads and load combinations the structure or component is expected to withstand;
(d)   Protection against the effects of internal hazards;
(e)   Protection against the effects of external hazards;
(f)   Design limits and acceptance criteria applicable to the design of structures, systems and components;
(g)   Reliability;
(h)   Provisions against common cause failures within a system and between systems belonging to different levels of defence in depth;
(i)   Safety classification;
(j)   Environmental conditions for qualification;
(k)   Monitoring and control capabilities;
(l)   Materials;
(m)   Provisions for testing, inspection, maintenance and decommissioning.


SAFETY FUNCTIONS

3.8.   The safety functions to be fulfilled by each system and the contribution of each major component should be described in a level of detail sufficient for correct safety classification.

## POSTULATED INITIATING EVENTS

3.9.   Paragraphs 3.10–3.12 provide recommendations on meeting Requirement 16 of SSR-2/1 (Rev. 1) [1].

3.10. From the list of postulated initiating events established for the design of the plant, those events that affect the design of the reactor coolant system and associated systems should be identified and categorized on the basis of their estimated frequency of occurrence (see Ref. [7] on the categorization of postulated initiating events).

3.11. For each of the conditions caused by the postulated initiating events, a list of the reactor coolant system and associated systems that are necessary to bring the plant to a safe and stable shutdown condition should be established.

3.12. Bounding conditions caused by the postulated initiating events should be determined in order to define the capabilities and performance of the reactor coolant system and associated systems and related equipment.

## INTERNAL HAZARDS

3.13. Paragraphs 3.14–3.17 provide recommendations on meeting Requirement 17 and para. 5.16 of SSR-2/1 (Rev. 1) [1] in relation to internal hazards. The recommendations provided in IAEA Safety Standards Series Nos NS-G-1.7, Protection against Internal Fires and Explosions in the Design of Nuclear Power Plants [8], and NS-G-1.11, Protection against Internal Hazards other than Fires and Explosions in the Design of Nuclear Power Plants [9], should be followed to identify internal hazards to be considered in the design of the reactor coolant system and associated systems.

3.14. The screening process used for identifying internal hazards should be documented in accordance with the management system. Structures, systems and components important to safety (e.g. for the safe shutdown of the reactor or for the mitigation of the consequences of an accident) should be protected against the effects of internal hazards. This protection should also consider the consequences of the effects of the failure of non-protected structures, systems and components on protected structures, systems and components.

3.15. The plant layout and the means for protection against internal hazards should be adequate to ensure that the response of the reactor coolant system and associated systems, as described in the analysis of the postulated initiating events, remains valid even when subjected to the effects of the hazard.

3.16. The plant layout and the means for protection of the redundancy provisions of the safety systems should be adequate to provide assurance that an internal hazard cannot represent a common cause failure for the total loss of a safety function to be fulfilled by the reactor coolant system and associated systems.

3.17. The design methods, as well as the design and construction codes and standards used, should provide adequate margins to avoid cliff edge effects in the event of an increase in the severity of the internal hazards.


EXTERNAL HAZARDS

3.18. Paragraphs 3.19–3.26 provide recommendations on meeting Requirement 17 and paras 5.17–5.21A of SSR-2/1 (Rev. 1) [1] in relation to external hazards. The recommendations provided in the following IAEA safety standards should also be considered to understand the general concepts, ensure identification of the relevant external hazards and protect the reactor coolant system and associated systems against the effects of these hazards: NS-G-1.5, External Events Excluding Earthquakes in the Design of Nuclear Power Plants [10], with regard to hazards other than seismic; and both NS-G-2.13, Evaluation of Seismic Safety for Existing Nuclear Installations [11], and NS-G-1.6, Seismic Design and Qualification for Nuclear Power Plants [12], with regard to seismic hazards.

3.19. With regard to the effects of external hazards, protection should be applied to the extent possible to prevent damage to the reactor coolant system and associated systems that are important to safety (e.g. systems to shut down the reactor and to mitigate the consequences of an accident). Protection can rely on an adequate layout of the plant and on protection measures for the buildings at the site. When protection measures are not effective, structures, systems and components should be designed to withstand the hazard loads and the loads associated with likely combinations of hazards.

3.20. The design of the components of the reactor coolant system should be such that the effects of the external hazards identified in the site evaluation cannot initiate an accident.

3.21. For each relevant external hazard or likely combination of hazards, components whose operability or integrity is necessary during or after the hazard induced event should be identified and specified in the design basis of the components.

3.22. Structures, systems and components of the reactor coolant system and associated systems should be assigned to appropriate seismic categories in accordance with the recommendations provided in NS-G-1.6 [12]. Components forming the reactor coolant pressure boundary, the secondary envelope of the steam generators (for PWRs and PHWRs), and the safety systems designed to mitigate the consequences of design basis accidents should be designed to withstand seismic level 2 (SL-2) seismic loads, in accordance with the recommendations provided in IAEA Safety Standards Series No. SSG-9, Seismic Hazards in Site Evaluation for Nuclear Installations [13].

3.23. The design methods, as well as the design and construction codes and standards used, should provide adequate margins to avoid cliff edge effects in the event of an increase in the severity of the external hazards.

3.24. The reactor coolant system and associated systems, and the structures and components ultimately necessary to prevent an early radioactive release or a large radioactive release (if any) should be identified. For all such items, the integrity and operability (where relevant) should be preserved in the event of natural hazards causing loads exceeding those determined from the hazard evaluation for the site. The boundary conditions selected for the design or assessment should be justified.

3.25. For external hazards, short term actions necessary to preserve the integrity of the reactor coolant pressure boundary and to prevent conditions from escalating to design extension conditions with core melting are required to be accomplished by systems available at the site (see para. 5.17 of SSR-2/1 (Rev. 1) [1]).

3.26. The capability for adequate core cooling should be such that it will remain operational for longer than the time at which off-site support services are credited.


ACCIDENT CONDITIONS

3.27. Accident conditions relevant for the design of the reactor coolant system and associated systems are those conditions with the potential to cause excessive mechanical loads to the reactor coolant system and associated systems, or those conditions for which the cooling of the fuel and the reactor shutdown would no longer be completed by the systems designed for operational states.

3.28. Accident conditions should be used as inputs for determining capabilities, loads and environmental conditions in the design of the reactor coolant system and associated systems. Accident conditions to be considered for the reactor coolant system and associated systems include the following:

(a) Loss of coolant accidents;
(b) Reactor coolant leakages to the secondary side (for PWRs and PHWRs);
(c) Main steam line break or steam generator feedwater pipe break (for PWRs and PHWRs);
(d) Loss of residual heat removal in shutdown conditions;
(e) Reactivity and power distribution anomalies.

3.29. Paragraphs 3.30–3.32 provide recommendations on meeting Requirement 18 of SSR-2/1 (Rev. 1) [1].

3.30. The computer codes and engineering rules that are used for the design of the reactor coolant system and associated systems should be documented, validated and, in the case of new codes, developed in accordance with up to date knowledge and recognized standards for management systems. Users of the codes should be qualified and trained with respect to the validation and application of the codes and to the assumptions made in the models in the codes.

3.31. The calculation of boundary conditions for design basis accidents and design extension conditions should be adequately documented, indicating the assumptions used for the evaluation of parameters, as well as the engineering criteria and the computer codes used.

3.32. Computer codes should not be used beyond their identified and documented domain of validation.

**Design basis accidents**

3.33. Paragraphs 3.34 and 3.35 provide recommendations on meeting Requirements 19 and 25 of SSR-2/1 (Rev. 1) [1].

3.34. Design basis accidents should be identified and the behaviour of the reactor coolant system should be analysed in order to specify the adequate performance of the safety systems.

3.35. To ensure adequate performance of the reactor coolant system and associated systems, conditions associated with design basis accidents should be calculated taking into account the least favourable initial conditions and equipment performance, as well as the single failure that has the greatest impact on the performance of the safety systems. Care should be taken when introducing adequate conservatism on account of the following:

(a) For the same event, an approach that is considered conservative for designing one specific system could be non-conservative for another. Consequently, various analyses should be performed for the different cases.

(b) Making overconservative assumptions could lead to the imposition of excessive stresses on components and structures.

**Design extension conditions without significant fuel degradation**

3.36. Paragraphs 3.37–3.42 provide recommendations on meeting Requirement 20 of SSR-2/1 (Rev. 1) [1] for design extension conditions without significant fuel degradation.

3.37. Relevant design extension conditions should be identified on the basis of engineering judgement as well as deterministic and probabilistic assessment.

3.38. As typical examples, the following three types of design extension condition should be considered:

(a) Events with a very low frequency of occurrence that could lead to situations beyond the capability of the safety systems to meet the acceptance criteria relevant for design basis accidents;

(b) Multiple failures (e.g. common cause failures in redundancies) that prevent the safety systems from performing their intended function to control the postulated initiating event;

(c) Multiple failures that cause the loss of the heat transfer chain to the ultimate heat sink during normal operation.

3.39. As multiple failures are likely to be caused by the occurrence of dependent failures that might lead to the failure of the safety systems, an analysis of dependences between redundant trains of safety systems or between diverse installed capabilities to shut down the reactor, to remove residual heat from the core and to transfer residual heat to the ultimate heat sink should be conducted to identify relevant candidates for design extension conditions.

3.40. Design extension conditions without significant fuel degradation should be identified and used to establish the design bases of systems necessary to prevent postulated sequences with multiple failures from escalating to design extension conditions with core melting. Examples of design extension conditions that might apply include:

(a)     Station blackout;
(b)     Anticipated transient without scram (for PWRs and BWRs);
(c)     Total loss of the feedwater systems (for PWRs and PHWRs);
(d)     A small loss of coolant accident with failures in the emergency core cooling system;
(e)     Loss of the residual heat transfer systems to the ultimate heat sink;
(f)     Loss of the ultimate heat sink.

3.41. For determining the necessary performance of the reactor coolant system and associated systems necessary in design extension conditions, the design extension conditions may be calculated with less conservative rules than those used for design basis accidents, provided that the margins are still sufficient to cover uncertainties. Performing sensitivity analyses could also be useful to identify which key parameters present uncertainties that should be considered in the design.

3.42. To the extent possible, the mitigation of the consequences of design extension conditions should be accomplished by permanent systems for cooling. Short term actions should be implemented by permanent equipment.


DESIGN LIMITS AND ACCEPTANCE CRITERIA

3.43. Paragraph 3.44 provides recommendations on meeting Requirements 15 and 28 of SSR-2/1 (Rev. 1) [1].

3.44. The performance of the reactor coolant system and associated systems should be specified to meet a well defined and accepted[2] set of design limits and criteria, in accordance with the following recommendations:

(a)     Components of the reactor coolant system should be designed so that the relevant limits for process parameters and stresses and the cumulative usage factor — to ensure the necessary integrity and operability of these components — are not exceeded.

---

[2] 'Well defined and accepted' generally means either widely accepted by Member State regulatory bodies or proposed by international organizations.

(b)    Associated systems should be designed so that the relevant design limits and criteria for fuel are not exceeded.

(c)    Associated systems should be designed so as not to cause unacceptable stresses on the reactor coolant pressure boundary.

3.45. Design limits and criteria are required to be specified for each plant state (see Requirement 15 of SSR-2/1 (Rev. 1) [1]).


## RELIABILITY

3.46. Paragraphs 3.47–3.56 provide recommendations on meeting Requirements 21–26, 29 and 30 of SSR-2/1 (Rev. 1) [1].

3.47. To achieve the necessary reliability of the reactor coolant system and associated systems to control the reactivity of the core, to maintain sufficient inventory in the reactor coolant system, to remove residual heat from the core and to transfer residual heat to the ultimate heat sink, the following factors should be considered:

(a)    Safety classification and the associated engineering requirements for design and manufacturing;

(b)    Design criteria relevant for the systems (e.g. number of redundant trains, seismic qualification, qualification to harsh environmental conditions, and power supplies);

(c)    Prevention of common cause failures by the implementation of suitable measures such as diversity, physical separation and functional independence;

(d)    Layout provisions to protect the reactor coolant system and associated systems against the effects of internal and external hazards;

(e)    Periodic testing and inspection;

(f)    Ageing effects;

(g)    Maintenance;

(h)    Use of equipment designed for fail-safe behaviour.

**Systems designed to cope with design basis accidents**

3.48. Shutting down the reactor, cooling the core, controlling core reactivity, residual heat removal and transfer to the ultimate heat sink in the event of design basis accidents should all be possible despite consequential failures caused by the postulated initiating event and a single failure postulated in any system necessary to fulfil a safety function. The unavailability of systems due to maintenance or repair should also be considered.

3.49. Systems that maintain the reactor in a safe state in the long term should be designed to fulfil their function despite a single failure postulated in any of those systems (either an active failure or a passive failure: see para. 5.40 of SSR-2/1 (Rev. 1) [1]). Some component failures might not need to be postulated (e.g. some passive failures) if this is duly justified.

3.50. The on-site power source (i.e. the emergency diesel generator and/or batteries) should have adequate capability to supply power to electrical equipment to be operated in design basis accidents for shutting down the reactor, cooling the core, removing and transferring residual heat to the ultimate heat sink, and maintaining the reactor in a safe state in the long term. More detailed recommendations are provided in IAEA Safety Standards Series No. SSG-34, Design of Electrical Power Systems for Nuclear Power Plants [14].

3.51. Vulnerabilities to common cause failures between the redundancies of the safety systems should be identified, and design and layout provisions should be implemented to make the redundancies independent as far as is practicable. In particular, adequate physical separation should be implemented between the redundant trains of the safety systems to prevent or minimize common cause failure due to the effects of hazards considered for design.

3.52. Recommendations relating to the reliability of the reactor coolant system and associated systems with regard to the effects of internal hazards, external hazards and environmental conditions are addressed in paras 3.14–3.17, 3.19–3.26 and 3.68–3.75, respectively.

**Safety features for design extension conditions without significant fuel degradation**

3.53. A reliability analysis of the safety systems designed to remove residual heat and to transfer residual heat to the ultimate heat sink should be conducted to identify the need for additional safety features to reinforce the prevention of core melting.

3.54. The more likely combinations of postulated initiating events and common cause failures between the redundancies of the safety systems should be analysed. If the consequences exceed the limits given for design basis accidents, the reliability of the safety systems should be improved (e.g. vulnerabilities to common cause failures should be removed) or additional design features should be implemented to prevent such events from escalating to an accident with core melting. The additional features for residual heat removal and residual heat transfer to the ultimate heat sink should be designed and installed such that they will be unlikely to fail as a result of the same cause.

3.55. The additional safety features should have a reliability that is sufficient to meet the core damage frequency criterion.

3.56. The recommendations in paras 3.48–3.52 should also be applied in respect of design extension conditions, taking into account that meeting the single failure criterion is not necessary and that the additional safety features for design extension conditions are supplied by the alternate AC power source and batteries.


DEFENCE IN DEPTH

3.57. Paragraphs 3.58–3.61 provide recommendations on meeting Requirement 7 of SSR-2/1 (Rev. 1) [1].

3.58. Alternative means to shut down the reactor or to maintain subcriticality, and to accomplish residual heat removal and residual heat transfer to the ultimate heat sink in different plant states should be implemented in accordance with the concept of defence in depth.

3.59. Vulnerabilities to common cause failure between these alternative means should be identified and the consequences of such a failure should be assessed. The vulnerabilities to common cause failure should be removed to the extent possible where an escalation to an accident with core melting would be the consequence of such a failure.

3.60. The independence implemented between instrumentation and control systems, or in the support systems necessary for the actuation and operation of the instrumentation and control systems, should not be compromised by common cause failure.

3.61. The instrumentation architecture supporting the actuation of the reactor coolant system and associated systems that is designed as safety systems should be independent, as far as is practicable, from instrumentation for monitoring the plant status.

## SAFETY CLASSIFICATION

3.62. Paragraphs 3.63–3.66 provide recommendations on meeting Requirement 22 of SSR-2/1 (Rev. 1) [1]. The recommendations provided in IAEA Safety Standards Series No. SSG-30, Safety Classification of Structures, Systems and Components in Nuclear Power Plants [15], should also be considered.

3.63. The consequences of a failure of a structure, system or component should be considered in terms of both the fulfilment of the safety function and the prevention of a radioactive release. For structures, systems and components for which both these factors are relevant, the safety class and the associated quality requirements that are necessary to achieve the expected reliability should be defined with due account taken of these two factors. For structures, systems and components that do not contain radioactive material, the safety class and the quality requirements should be directly derived from the consequences of the safety function not being fulfilled.

3.64. The safety classification should be established in a consistent manner such that all systems necessary for the fulfilment of one safety function (including the associated support service systems) are assigned to the same class, or else a justification for assigning a different class should be provided.

3.65. In accordance with Requirement 9 of SSR-2/1 (Rev. 1) [1], pressure retaining equipment that is safety classified is required to be designed and manufactured in accordance with proven codes and standards widely used by the nuclear industry. More detailed recommendations are provided in IAEA Safety Standards Series Nos NS-G-2.2, Operational Limits and Conditions and Operating Procedures for Nuclear Power Plants [16], SSG-52, Design of the Reactor Core for Nuclear Power Plants [17], and SSG-63, Design of Fuel Handling and Storage Systems for Nuclear Power Plants [18]. The engineering design and manufacturing rules applicable to each individual component should be selected with due account taken of the consequences of failure, in terms of both the fulfilment of the safety function and the prevention of a radioactive release.

3.66. With regard to implementing the safety classification described in SSG-30 [15]:

(a)   Systems that are designed to prevent dose limits being exceeded in the event of a design basis accident should be assigned to safety class 1, or these systems may be assigned to safety class 2 if they are necessary to bring the reactor to a safe state.

(b)   Systems implemented to provide a backup to the safety features for design extension conditions should be assigned to at least safety class 2. For the categorization of a safety function as safety category 2, para. 3.15 of SSG-30 [15] includes: "Any function that is designed to provide a backup of a function categorized in safety category 1 and that is required to control design extension conditions without core melt."

(c)   Systems designed to keep the key reactor parameters (e.g. pressure, temperature, pressurizer water level and steam generator water level) within the ranges specified for normal operation should be assigned to at least safety class 3.

(d)   Systems designed for normal operation and whose failure would not lead to radiological consequences exceeding the authorized limits specified for operational conditions need not be safety classified.


## ENVIRONMENTAL QUALIFICATION OF ITEMS IMPORTANT TO SAFETY

3.67. Paragraphs 3.68–3.75 provide recommendations on meeting Requirement 30 of SSR-2/1 (Rev. 1) [1]. The recommendations provided in IAEA Safety Standards Series Nos SSG-48, Ageing Management and Development of a Programme for Long Term Operation of Nuclear Power Plants [19], and NS-G-1.6 [12] should also be considered.

3.68. The components and instrumentation for the reactor coolant system and associated systems are required to be qualified to perform their functions in the entire range of environmental conditions that might prevail prior to or during their operation, or should otherwise be adequately protected from those environmental conditions (see Requirement 30 of SSR-2/1 (Rev. 1) [1]).

3.69. The relevant environmental conditions that might prevail prior to, during and following an accident, and the ageing of structures, systems and components throughout the lifetime of the plant are all required to be taken into consideration in the environmental qualification (see Requirement 30 of SSR-2/1 (Rev. 1) [1]). Further recommendations on ageing management are given in SSG-48 [19].

3.70. Environmental qualification should be carried out by means of testing, analysis and the use of experience, or through a combination of these.

3.71. Environmental qualification should include the consideration of such factors as temperature, pressure, humidity and radiation levels. Margins and synergistic effects (in which the damage due to the superposition or combination of effects might exceed the total damage due to the effects separately) should also be considered. In cases where synergistic effects are possible, materials should be qualified for the most severe effect or the most severe combination or sequence of effects.

3.72. Techniques to accelerate the testing for ageing and qualification can be used, provided that there is adequate justification to do this.

3.73. For components subject to the effects of ageing degradation by various mechanisms, the design life and, if necessary, the replacement frequency should be established. In the qualification process for such components, samples should be aged to simulate the end of their design lives before being tested under relevant accident conditions.

3.74. Components that have been used for qualification testing should generally not be used subsequently for construction purposes.

3.75. Evidence of environmental qualification, the applicable parameters and the established qualification needs should be contained in (or referenced by) design documentation in an auditable form throughout the lifetime of the plant.


LOADS AND LOAD COMBINATIONS

3.76. The design basis of each component and structure of the reactor coolant system and associated systems should include — for each plant state and service condition — the loads and load combinations imposed by construction, lifting and environmental conditions inside the buildings, and the internal and external hazards for which stability, integrity, functionality and operability are necessary.

3.77. Loading conditions, loads and stresses should be calculated by applying adequate and accepted methodologies and rules to establish confidence in the robustness of the design and to provide adequate margins to cover uncertainties and avoid cliff edge effects, taking into account the following:

(a)　Uncertainties in process parameters;
(b)　Uncertainties in initial conditions and the performance of systems and components;
(c)　Uncertainties in models;

(d)     Structural tolerances;
(e)     Uncertainties in relation to the decay heat.

3.78. Loads should be identified and analysed, with account taken of the following:

(a)     The type of load (i.e. static and permanent loads, or transient and dynamic, global or local);
(b)     The timing of each load (to avoid the unrealistic superposition of load peaks if such peaks cannot occur coincidently).

3.79. Design basis loading conditions, including internal and external hazard loads, should be assigned to different categories that correspond to different plants states or service conditions[3] according to their estimated frequency of occurrence or in accordance with the requirements of accepted codes and national regulations.

3.80. Appropriate acceptance criteria (e.g. design pressure and temperature, and stress limits) to be met for ensuring integrity should be defined and should be appropriate to each load combination, with account taken of the load combination category.

3.81. The stress levels might be different for the different modes of failure (e.g. progressive deformation and fatigue, or excessive deformation and plastic instability). Protection against brittle fracture should be ensured, and the critical buckling stress should be considered if relevant for the component.

---

[3] The categories of service condition are defined as follows:
— Normal service conditions: Loading conditions to which the equipment might be subjected during normal operation including normal operating transients and startup and shutdown conditions.
— Upset conditions: Loading conditions to which the equipment might be subjected during transients resulting from the occurrence of a postulated initiating event categorized as an anticipated operational occurrence.
— Emergency conditions: Loading conditions to which the equipment might be subjected during transients resulting from the occurrence of a postulated initiating event categorized as an accident of low frequency.
— Faulted conditions: Loading conditions to which the equipment might be subjected during transients resulting from the occurrence of a postulated initiating event categorized as an accident of very low frequency.

3.82. Meeting the criteria given by internationally recognized codes and standards provides reasonable assurance that structures, systems and components are capable of performing their intended functions. When operability needs to be demonstrated, additional analyses or tests should be conducted.

3.83. Normal service conditions and upset conditions should be defined by modelling the plant response under realistic conditions.

3.84. Emergency conditions and faulted conditions should be defined with conservatism (e.g. by assuming unfavourable uncertainties in the initial conditions and in the performance of the systems, and not crediting the operational systems and controls when their operation is favourable).

3.85. Structures, systems and components that are necessary for the mitigation of the consequences of an accident should be designed to withstand the effects of natural phenomena in order to retain their capability to fulfil their intended safety functions. The design bases of these structures, systems and components should reflect appropriate combinations of the effects of operational states and accident conditions with the effects of natural phenomena.

3.86. Structures, systems and components that are designed to fulfil their functions in emergency conditions and faulted conditions should be designed to meet adequate[4] service limits, to ensure the necessary integrity and operability of these items while subjected to sustained loads resulting from the occurrence of the postulated initiating events for which they are designed to respond.

MATERIALS

3.87. Paragraphs 3.88–3.92 provide recommendations on meeting Requirement 47 of SSR-2/1 (Rev. 1) [1].

3.88. The materials used for the pressure retaining boundary of the reactor coolant system and associated systems should be specified with regard to chemical composition, microstructure, mechanical–thermal properties, heat treatment, manufacturing requirements and activation of materials, as applicable. The materials should be appropriately homogeneous and should be compatible with the coolant they contain, as well as with joining materials (e.g. welding materials)

---

[4] Meeting the stress limit given by codes for emergency conditions or faulted conditions is generally not considered adequate by the regulatory body.

and with adjoining components or materials, such as sliding surfaces, spindles and stuffing boxes (packing boxes), overlay or radiolysis products.

3.89. The specifications for welding materials used for the manufacture or repair of components should also be established so that the welds have sufficient strength and toughness.

3.90. Materials specified for the reactor coolant system and associated systems should comply with the applicable provisions of the code used, including the following properties and characteristics:

(a)   Resistance to heat loads;
(b)   Strength, creep and fatigue properties;
(c)   Corrosion and erosion related properties, including resistance to stress corrosion cracking;
(d)   Resistance to effects of irradiation;
(e)   Resistance to thermal embrittlement;
(f)   Resistance to hydrogen embrittlement;
(g)   Ductility characteristics;
(h)   Fracture toughness characteristics (including both brittle and ductile fracture toughness);
(i)   Ease of fabrication (including weldability).

3.91. As stated in para. 6.70 of SSR-2/1 (Rev. 1) [1], materials used in the manufacture of structures, systems and components of the reactor coolant system and associated systems "shall be selected to minimize activation of the material as far as is reasonably practicable."

3.92. The materials selected should be suitable for the service conditions expected throughout the lifetime of the nuclear power plant and for all operational states and accident conditions. They should be qualified by means of analysis, testing and the feedback and analysis of operating experience, or a combination of these.

**Materials in contact with radioactive fluids**

3.93. Materials should be highly resistant to all corrosion phenomena in operational states, including any deterioration due to corrosion by the fluid and the abrasive effects of suspended solids.

3.94. In accordance with para. 4.20 of SSR-2/1 (Rev. 1) [1], it is required that the materials used will facilitate decontamination.

**Materials exposed to high neutron flux**

3.95. The choice of materials used in this application should take into consideration the following effects:

(a)  Embrittlement due to neutron irradiation;
(b)  Irradiation assisted stress corrosion cracking;
(c)  Swelling due to neutron irradiation;
(d)  Neutron activation;
(e)  Irradiation creep.

3.96. With regard to the risk of embrittlement of the reactor pressure vessel, a surveillance programme should be established on the basis of tests conducted on samples of the materials used for the manufacture of the reactor pressure vessel. These samples should be installed in the reactor pressure vessel and removed on a scheduled basis. These samples should then be subjected to mechanical testing, including tensile strength and Charpy impact testing or fracture toughness testing. Other samples should be analysed to measure the irradiation fluence to which the wall of the reactor pressure vessel and the samples are being exposed. Acceptance criteria should be specified for all the tests performed.

## MANUFACTURING AND INSTALLATION

3.97. Paragraph 3.98 provides recommendations on meeting Requirement 11 of SSR-2/1 (Rev. 1) [1].

3.98. In accordance with Requirement 11 of SSR-2/1 (Rev. 1) [1], pressure retaining components:

> **"shall be designed so that they can be manufactured, constructed, assembled, installed and erected in accordance with established processes that ensure the achievement of the design specifications and the required level of safety."**

A management system should be established for the manufacturing process, including identification and traceability of materials, welding, handling and storage of manufactured components. More detailed criteria are provided in the codes and standards selected for the manufacturing.

## CALIBRATION, TESTING, MAINTENANCE, REPAIR, REPLACEMENT, INSPECTION AND MONITORING

3.99. Paragraphs 3.100–3.115 provide recommendations on meeting Requirement 29 of SSR-2/1 (Rev. 1) [1].

3.100. In accordance with Requirement 29 of SSR-2/1 (Rev. 1) [1], a range of measures is required to be taken to ensure that structures, systems and components important to safety will keep their capability to perform their intended function over their service life. Inspection and periodic testing are good practices that help to meet this requirement.

3.101. The design should establish a technical basis for structures, systems and components that require in-service inspection, examination, testing, maintenance and monitoring.

3.102. The design should incorporate provisions to facilitate examination, testing, in-service inspection, maintenance, repair and monitoring to be carried out during the construction, commissioning and operation stages. Further recommendations are provided in IAEA Safety Standards Series No. NS-G-2.6, Maintenance, Surveillance and In-service Inspection in Nuclear Power Plants [20].

3.103. Structures, systems and components important to safety should be designed and located to make surveillance and maintenance simple, to permit timely access, and in case of failure, to allow diagnosis and repair and to minimize risks to maintenance personnel.

3.104. The development of strategies and programmes to address examination, testing, in-service inspection, maintenance and monitoring is an essential aspect of the design of the reactor coolant system and associated systems. The strategies and programmes to be implemented should take into account human factors engineering in order to facilitate the efficient conduct of activities and to minimize the contribution of human error.

3.105. If the plant design contains safety equipment that cannot be tested in situ (e.g. explosively actuated valves), an appropriate surveillance programme should be implemented that includes pre-service and in-service provisions.

3.106. Non-destructive examinations should be defined and conducted on welds and weld claddings to ensure the acceptability of their structural integrity on the basis of predefined acceptance criteria for each type of non-destructive method.

Personnel, equipment and procedures should be qualified prior to performing the non-destructive examination.

**Pre-service and in-service inspection of the reactor coolant system**

3.107.   The components of the reactor coolant pressure boundary should be designed, manufactured and installed in a manner that permits adequate inspections and tests of the boundary, support structures and components throughout the lifetime of the plant.

3.108.   The design should allow access to any part of the reactor coolant system that has to be inspected throughout the lifetime of the plant, in particular to welds. Specific areas subject to cyclic loads and neutron irradiation should be identified at the design stage and should be specifically monitored in order to confirm that no unacceptable damage occurs owing to ageing effects, thermal fatigue or neutron irradiation.

3.109.   Methods and criteria provided by relevant national and international codes and standards might be used for pre-service inspection and for in-service inspection.

*Pre-service inspection and testing*

3.110.   Prior to the start of operation, a pre-service inspection programme should be developed and implemented.

3.111.   The reactor pressure vessel and the reactor coolant pressure boundary should be subject to examinations, inspections and tests to ensure that the vessel and components have been correctly manufactured and installed. These include the following examinations and tests:

(a)   Hydrostatic pressure test by the manufacturer of the reactor pressure vessel and all other vessels, valve bodies and casings prior to installation.
(b)   Non-destructive examinations of the reactor pressure vessel and reactor coolant pressure boundary welds and other representative areas, utilizing volumetric ('through wall') and surface examinations. These examinations are important to establish the baseline condition to be used for comparison with the in-service inspection results.
(c)   Hydrostatic test (in accordance with the design and the manufacturing code) of the reactor pressure vessel and reactor coolant system once installation is complete.

During the performance of the pre-service inspection programme, design features should be identified to facilitate and simplify the implementation of the in-service inspection programme during operation. This should take into account that many areas will not be easily accessible once operation commences. In such cases, adequate provisions should be made for the inspection of those areas to the extent practicable.

*In-service inspection and testing*

3.112. The design of the reactor pressure vessel and the reactor coolant pressure boundary should allow for volumetric examination of the entire volume of the welds as well as for surface examinations. For example, ultrasonic, eddy current or magnetic flux methods could be used for such examinations.

3.113. Welds of the reactor pressure vessel and the reactor coolant pressure boundary that cannot be inspected in-service should be limited to the extent possible, and analyses of the consequences of the failure of such welds should be performed.

3.114. The following should be considered in deriving the inspection criteria:

(a) The minimum detectable indication in non-destructive examinations;
(b) The expected crack growth and fracture toughness in operational states and in accident conditions;
(c) The sourcing of welded and base metal coupons representing relevant inspection areas of the reactor pressure vessel and other major components subject to recurrent ultrasonic testing (e.g. welded joints and base metal with cladding, bimetallic welds and nozzle areas) for ultrasonic testing calibration blocks;
(d) The maximum acceptable defect in operational states;
(e) The commissioning (code) hydrostatic pressure test;
(f) Periodic leak rate and hydrostatic tests;
(g) The periodic in-service inspection programme specified by relevant codes;
(h) All controls during the manufacturing process to be referenced and traceable for the operational lifetime.

**Inspection of steam generators**

3.115. The design of the steam generators should allow for inspection of the steam generator tubes over their entire length. The equipment and procedures for examination of the tubes should be capable of detecting and locating significant defects.

## OVERPRESSURE PROTECTION

3.116. Paragraphs 3.117–3.120 provide recommendations on meeting Requirement 48 of SSR-2/1 (Rev. 1) [1].

3.117. All pressure retaining components of the reactor coolant system and associated systems should be protected against overpressure conditions generated by component failures or by anticipated operational occurrences in order to ensure the integrity of the component in compliance with applicable proven codes and standards.

3.118. Overpressure protection devices should be installed as close as practicable to the component to be protected.

3.119. The discharge capacity should be sufficient to limit the pressure such that the stress limits for the service condition are met for each of the components to be protected.

3.120. The same code should be used for the design, manufacturing and overpressure analysis of a given component.

## LAYOUT

3.121. The design of the layout of the reactor coolant system and associated systems should take into account:

(a)	Radiation protection of site personnel;
(b)	Protection against the consequences of pipe failure (e.g. depressurization waves, pipe whip, flooding and high pressure jet impingement, including potential blast waves);
(c)	Protection against internal missiles;
(d)	Provisions for venting and draining the reactor coolant;
(e)	Provisions to avoid water stratification and accumulation of gases;
(f)	Provisions to avoid erosion;
(g)	Provisions to avoid water hammer;
(h)	Provisions for seismic events;

(i)    Provisions to minimize stresses in the piping (including a consideration of thermal expansion);

(j)    Provisions to facilitate testing, inspection, repair and replacement.


## RADIATION PROTECTION

3.122.  Paragraphs 3.123–3.125 provide recommendations on meeting Requirement 81 of SSR-2/1 (Rev. 1) [1]. Detailed recommendations for design measures for radiation protection are provided in IAEA Safety Standards Series No. NS-G-1.13, Radiation Protection Aspects of Design for Nuclear Power Plants [21].

3.123.  The design of the layout of the reactor coolant system and associated systems should allow for the inspection, maintenance, repair and replacement of components, taking into account the requirement to optimize protection and safety by keeping radiation risks as low as reasonably achievable (see para. 3.6(a) of SSR-2/1 (Rev. 1) [1]).

3.124.  Appropriate design provisions (e.g. shielding and remote control valves) should be implemented to enable the local actions necessary for the management of accidents to be undertaken without undue radiation exposure of operating personnel. Similar design provisions should be implemented to enable the recovery of systems necessary to maintain safe conditions in the long term while ensuring that protection and safety are optimized.

3.125.  The quantities of antimony, cobalt, silver and other easily activated elements in materials in contact with the reactor coolant should be minimized to reduce the activation of entrained corrosion products with radionuclides such as $^{124}$Sb, $^{60}$Co and $^{110}$Ag.


## COMBUSTIBLE GAS ACCUMULATION IN NORMAL OPERATION

3.126.  Design and layout provisions should be implemented to prevent the accumulation of combustible gases in the upper parts of components (e.g. the upper part of the reactor pressure vessel, pressurizer and safety valves) and piping.

## VENTING AND DRAINING

3.127.   Provision should be made for venting and draining the reactor coolant system and associated systems.

3.128.   Provisions should also be implemented for collecting and managing coolant inventories from leakages during normal operation. During reactor operation, leakages can occur from various components, including valve stems, valve seats, pump seals and gaskets.

## INTERFACES BETWEEN THE REACTOR COOLANT SYSTEM AND ASSOCIATED SYSTEMS

3.129.   Appropriate isolation devices should be provided for connections between systems or components belonging to different safety classes (see SSG-30 [15]). These devices should prevent situations in which the failure of a system or a component could cause the loss of the safety function of the system or component in the higher safety class, and should limit the release of radioactive material. The isolation device should be assigned to the same safety class as the system or component in the higher safety class to which it is connected.

3.130.   Structures interfacing with the reactor coolant system and associated systems should be considered as items important to safety and should be designed accordingly to ensure their integrity and performance. Such structures include the following:

(a)   Snubbers and their anchors;
(b)   Pipe whip restraints;
(c)   Building penetrations;
(d)   Protective structures (e.g. barriers and shields).

## CONTAINMENT ISOLATION

3.131.   Paragraph 3.132 provides recommendations on meeting Requirement 56 of SSR-2/1 (Rev. 1) [1].

3.132.   Piping that penetrates the primary containment wall(s) is required to be provided with adequate isolation devices (see para. 6.22 of SSR-2/1 (Rev. 1) [1]). The piping run between isolation valves should be designed in accordance with the recommendations provided in IAEA Safety Standards Series No. SSG-53, Design of the Reactor Containment and Associated Systems for Nuclear Power Plants [22].


INSTRUMENTATION

3.133.   The reactor coolant system and associated systems should be provided with adequate instrumentation for the following purposes:

(a)   Monitoring of the process parameters (e.g. pressure, temperature, water level and flow rate) that indicate whether the system or component is being operated within the range specified for its normal operation;
(b)   Early detection of abnormal operating conditions;
(c)   Automatic operation of systems necessary for the mitigation of the consequences of an accident;
(d)   Providing the main control room and the technical support centre with appropriate and reliable information for accident management;
(e)   Periodic testing of systems and components;
(f)   Supporting an understanding of the maintenance state of structures, systems and components.

3.134.   The consequences of sharing sensors that have different purposes should be assessed in order to preserve adequate independence of the different levels of defence in depth. The following recommendations should be implemented to the extent possible:

(a)   Sensors for the automatic actuation of the operation of systems and for accident monitoring of the plant should not be shared.
(b)   The same installed sensors should not be shared with the automatic actuation of the reactor shutdown system or of the operation of the safety systems or with the actuation of the safety features implemented to reinforce the prevention of accidents with core melting.

3.135.   Instrument sensing lines should be designed such that the measurement parameters (e.g. magnitude, frequency, response time and chemical characteristics) are not distorted.

3.136.   Potential leakage of radioactive material from and into the reactor coolant system and associated systems should be monitored. Further recommendations are provided in IAEA Safety Standards Series No. SSG-39, Design of Instrumentation and Control Systems for Nuclear Power Plants [23].

## MULTIPLE UNITS AT A SITE

3.137.   As stated in Requirement 33 of SSR-2/1 (Rev. 1) [1]: **"Each unit of a multiple unit nuclear power plant shall have its own safety systems and shall have its own safety features for design extension conditions."**

## CODES AND STANDARDS

3.138.   Paragraph 3.139 provides recommendations on meeting Requirement 9 and paras 4.14–4.16 of SSR-2/1 (Rev. 1) [1].

3.139.   Proven and widely accepted codes and standards are required to be used for the design of the reactor coolant system and associated systems. The selected codes and standards should be applicable to the particular design and should form an integrated and comprehensive set of standards and criteria. For design and construction, the latest editions of the applicable codes and standards should preferably be considered. As stated in para. 4.16 of SSR-2/1 (Rev. 1) [1]:

> "Where an unproven design or feature is introduced or where there is a departure from an established engineering practice, safety shall be demonstrated by means of appropriate supporting research programmes, performance tests with specific acceptance criteria or the examination of operating experience from other relevant applications. The new design or feature or new practice shall also be adequately tested to the extent practicable before being brought into service, and shall be monitored in service to verify that the behaviour of the plant is as expected."

3.140.   Paragraph 4.6 of SSG-30 [15] states:

> "The licensee or applicant should provide and justify the correspondence between the safety class and the associated engineering design and manufacturing rules, including the codes and/or standards that apply to each SSC [structure, system and component]."

Codes and standards have been developed by various national and international organizations, covering areas such as:

(a)  Materials;
(b)  Manufacturing (e.g. welding) and construction;
(c)  Civil structures;
(d)  Pressure vessels and pipes;
(e)  Instrumentation and control;
(f)  Environmental and seismic qualification;
(g)  Pre-service and in-service inspection and testing;
(h)  The management system;
(i)  Fire protection.


USE OF PROBABILISTIC ANALYSES IN DESIGN

3.141.  Paragraphs 3.142 and 3.143 provide recommendations on meeting para. 5.76 of SSR-2/1 (Rev. 1) [1].

3.142.  Probabilistic analyses should be combined with a deterministic approach in confirming the reliability of the reactor coolant system and associated systems in terms of preventing significant fuel damage, and for identifying the more likely common cause failures and multiple failures that could be considered as initiators of design extension conditions.

3.143.  The use of probabilistic analyses should be part of the process to select optimal design options and to judge their effectiveness.


# 4. ULTIMATE HEAT SINK AND RESIDUAL HEAT TRANSFER SYSTEMS

4.1.  This section provides recommendations on meeting Requirement 53 of SSR-2/1 (Rev. 1) [1] with regard to the systems designed to transfer residual heat from the different decay heat removal systems to the ultimate heat sink.

4.2. The ultimate heat sink is the medium into which residual heat is discharged in the different plant states after shutdown of the reactor, and it normally consists of a large body of water or the atmosphere, or both. The body of water can be a sea, a river, a lake, a reservoir, groundwater or combinations of these, but in general access to inexhaustible, natural supplies of water is preferable to limited capacities. For an ultimate heat sink that relies on the atmosphere, cooling towers or spray ponds, with their associated structures and systems, are the usual equipment designed to transfer heat to the atmosphere. Some passive reactor plant designs also rely exclusively on the atmosphere for dissipating reactor decay heat immediately following plant transients and accident conditions. The medium used as a receptor for the decay heat can also be used as a source of cooling for turbine condensers during power operation; however, the associated heat transfer systems are out of the scope of this Safety Guide.

4.3. For a site with multiple units, the items important to safety designed as interfaces with the ultimate heat sink medium should be specific to each unit.

4.4. The capacity of the ultimate heat sink should be adequate to absorb decay heat from all the different reactors and spent fuel pools at the site. This capacity should be designed taking into consideration that all units could be in accident conditions simultaneously.

4.5. The reliability and capacity of the ultimate heat sink should be ensured for both the short term and the long term, taking into account all the relevant heat loads generated during normal shutdown modes, anticipated operational occurrences and accident conditions, the rates of heat rejection during those conditions and relevant regulations pertaining to environmental protection.

4.6. The short term and long term capacity of the ultimate heat sink should be preferably achieved by the use of inexhaustible, natural supplies of water or the atmosphere. Where access to an inexhaustible supply of water or the atmosphere at the site is not available:

(a)　The capacity of the ultimate heat sink should be ensured by an adequate amount of water always being available at the site. This capacity should be adequate to absorb all heat loads generated at the site until the heat sink can be replenished.[5] Account should be taken of factors that could delay the replenishment process. Such factors include evaporation, human induced events, natural hazards, accident conditions at the plant, availability of interconnections and the complexity of the procedures for replenishment.

(b)    A minimum amount of water, including a margin for uncertainties, should be immediately available to bring reactors to the safe shutdown state in the event of any postulated initiating event. For each unit, this minimum quantity[6] should already be stored in the basins of the cooling towers or spray pounds dedicated to the unit.

(c)    Beyond this minimum capacity, the additional water needed prior to replenishment could be stored in an on-site reservoir, with the possibility to transfer this water from the on-site reservoir to the ultimate heat sink. This transfer system should be considered as a support system to help to fulfil the safety function of the ultimate heat sink and should be safety classified accordingly.

(d)    In terms of the long term capacity of the ultimate heat sink, the make-up systems to replenish on-site reservoirs should be permanently installed and should be designed with an adequate rate to meet the long term heat removal capacity.

4.7.    To fulfil the design objectives in terms of capacity and reliability and to apply the concept of defence in depth (see Requirement 7 of SSR-2/1 (Rev. 1) [1]), the use of a different ultimate heat sink or different access to the ultimate heat sink might be necessary (see para. 6.19A of SSR-2/1 (Rev. 1) [1]).

4.8.    Structures associated with the ultimate heat sink should be designed to withstand the loads caused by the hazards derived from the hazard evaluation for the site. Recommendations on the consideration of external events (e.g. extreme temperatures and conditions, frazil ice, ice cover, floods, tsunamis, high winds, biological phenomena, collisions with floating bodies, clogging, low water levels, sand and sludge silting, and events involving hydrocarbons) in the design of such structures are provided in NS-G-1.5 [10].

4.9.    The provisions ensuring the effectiveness and availability of the ultimate heat sink with regard to the site's natural hazards should be designed with adequate

---

[5] In some States, the quantity of water that is immediately available (including water stored at the site in tanks or reservoirs) is thirty days, unless a shorter time period can be justified by conservative analysis.

[6] In some States, this quantity of water is designed to ensure heat removal capability for three days.

margins to cope with levels of natural hazards exceeding those derived from the hazard evaluation for the site (see para. 5.21A of SSR-2/1 (Rev. 1) [1]).

4.10. In determining the necessary capacity of the ultimate heat sink, design basis environmental parameters should be defined with account taken of the time periods during which those conditions are assumed to exist (see NS-G-1.5 [10]).

4.11. The effectiveness of the ultimate heat sink should not be unduly affected by short term variations in the environmental parameters.

4.12. The design basis environmental parameters should include the water temperature of the ultimate heat sink for 'once-through' water cooling systems and the dry bulb temperature of the air for dry cooling towers. Both wet bulb and dry bulb temperatures are necessary environmental parameters for wet cooling towers, cooling ponds and spray ponds and other heat transfer systems that use evaporative cooling.

4.13. It should be ensured that the capability for heat load rejection is maintained following any interruption of power generation or loss of operability of normal heat removal systems.

4.14. The ultimate heat sink should be designed to be capable of absorbing the relevant heat loads at the maximum peak heat rejection rate for the different plant states, with the time dependent behaviour of the individual heat loads taken into account.

4.15. In establishing the maximum heat rejection rate, the most severe combination of individual heat loads should be identified for all postulated initiating events for which the system is called upon to perform a normal operation or to fulfil a safety function.

4.16. In determining the capacities demanded of the ultimate heat sink and its directly associated heat transfer systems, the various heat sources and their time dependent behaviour should be precisely identified to ensure that the temperature of the coolant remains within specified limits. The heat loads that should be taken into consideration include the following:

(a)	Residual heat of the reactor coolant system;
(b)	Decay heat of the spent fuel with the storages at maximum capacity;
(c)	Heat generated by the operation of structures, systems and components to achieve and maintain a safe plant shutdown or to mitigate the consequences

of an accident (if heat produced by the components is transported by the residual heat transfer chain);

(d)     Heat from other accident related heat sources (e.g. chemical reactions).

4.17. In establishing the residual heat loads of the reactor (including decay heat, heat due to fission during shutdown, and the energy stored in the reactor coolant system and other operated heat removal systems or structures), it should be assumed that the fuel has been exposed to operation at power for a period of time that would produce the maximum decay heat load. The decay heat should be evaluated consistently with applicable standards.

4.18. The total heat load and the rejection rate of heat from spent fuel should be evaluated on the basis of the maximum number of spent fuel elements that can be stored at the site at any one time. Either the decay heat curves for the particular fuel (with appropriate individual post-shutdown times applied to the various fuel elements) or a conservative average post-shutdown time for all fuel elements should be used.

4.19. Accident conditions might produce additional sources of heat, such as the heat emanating from metal–water reactions of the fuel cladding or from other heat producing chemical reactions within the containment. If potential metal–water reactions are determined to be significant as an additional heat source, then they should be quantified as a function of time and included in the sizing criteria.

RESIDUAL HEAT TRANSFER SYSTEMS

4.20. The residual heat transfer chain includes the intermediate cooling systems and the cooling system directly associated with the ultimate heat sink. The intermediate cooling system is designed as a closed loop system that transfers heat from residual heat removal systems to the cooling system directly associated with the ultimate heat sink. The cooling system directly associated with the ultimate heat sink is an open loop system that takes water from the ultimate heat sink (pumping station), provides cooling to the intermediate cooling system and discharges transferred heat loads to the ultimate heat sink.

4.21. Paragraphs 4.22–4.27 provide recommendations on meeting Requirements 7 and 53 of SSR-2/1 (Rev. 1) [1].

4.22. All residual heat sources at the nuclear power plant should be considered in the design of the heat transfer systems.[7]

4.23. In accordance with the concept of defence in depth, the design should provide multiple means of transferring residual heat to the ultimate heat sink.

4.24. Where the heat removal system is not designed to operate the reactor coolant system in hot conditions, the residual heat removed by the secondary side can be directly released to the atmosphere, which constitutes a second ultimate heat sink (for PWRs and PHWRs in anticipated operational occurrences and accident conditions). In accordance with the concept of diversity, the operation of components necessary to feed and bleed the steam generators should not be directly dependent on the heat transfer chain.

4.25. To ensure the effectiveness of the application of the concept of defence in depth, the different heat transfer means provided should be independent to the extent practicable. In particular, a different and independent heat transfer chain should be implemented for accidents with core melting (see SSG-53 [22]).

4.26. The design and manufacture of the heat transfer chains and associated systems and components should apply the design recommendations derived from the safety class of these structures, systems and components, as determined on the basis of their safety significance.

4.27. Where an ultimate heat sink of limited capacity is provided, the choice of the heat transfer system with which it is directly associated might be dictated by the need to conserve the inventory of the ultimate heat sink; this would increase the time needed for make-up water to be available.

**Residual heat transfer in operational states**

4.28. Paragraphs 4.29–4.40 provide recommendations on meeting Requirement 51 of SSR-2/1 (Rev. 1) [1] and supplement the generic recommendations provided in Section 3.

4.29. Systems should be designed to transfer all heat loads that are generated from the control of primary coolant temperature in shutdown modes, and from the

---

[7] If the heat produced by the operation of some components is also removed and transported by those systems, the corresponding additional heat loads should be included.

control of the spent fuel pool temperature, within the temperature ranges specified for operational states.

4.30. The heat transfer should not be compromised by any single failure postulated for any component necessary for transferring residual heat to the ultimate heat sink.

4.31. Residual heat transfer should be possible in the event of a loss of off-site power.

4.32. The heat transfer chain should include an intermediate cooling system to prevent a leak of primary coolant being released into the ultimate heat sink.

4.33. Heat load transfer capabilities should be designed to be consistent with the requested performance of the residual heat removal system for the reactor and the spent fuel cooling system.

4.34. The heat transfer capacity for the spent fuel pool should be designed based on the maximum storage capacity of the pool, taking into account boundary conditions for the heat loads.

4.35. The heat transfer capacity should be designed to transfer heat loads generated during operational states based on the temperature of the ultimate heat sink remaining within the range defined for normal operation.

4.36. Residual heat transfer systems should be designed in compliance with the recommendations provided in paras 3.38, 3.40, 3.54 and 3.58 if these systems are also operated to transfer residual heat after a design basis accident (see paras 4.41–4.44). In such cases, a failure of equipment operated only during operational states should not propagate to equipment that is expected to operate during design basis accident conditions.

*Specific design aspects*

4.37. A monitoring system designed to detect radioactivity in the intermediate cooling system should be put in place.

4.38. The intermediate cooling system should be protected against overpressure caused by leakages occurring in heat exchangers that interface with cooling systems operated at higher pressure. In such cases, the intermediate cooling system should be designed to prevent primary coolant leakages outside the containment.

4.39. Cooling system pumps that are directly connected to the ultimate heat sink should be protected against debris and biofouling effects as follows:

(a)  A programme of monitoring for fouling of the heat exchangers and a cleaning programme should be implemented, with appropriate frequency, in order to limit the degradation of the heat removal capability of the system.
(b)  A programme of surveillance and control techniques should be implemented to significantly reduce the incidence of flow blockage problems due to biofouling or foreign objects.

4.40. The capabilities of the cooling system directly associated with the ultimate heat sink should be designed with account taken of the following:

(a)  Maximum heat rejection rate;
(b)  Environmental parameters for design (e.g. water and air temperatures, and relative humidity);
(c)  Supplies of coolant.

**Residual heat transfer for design basis accidents**

4.41. The design of the plant should include additional systems to transfer residual heat to the ultimate heat sink in the event of a design basis accident in cases where systems operated in normal shutdown conditions are not designed in compliance with engineering design requirements applicable to safety systems.

4.42. The heat transfer capacity should be designed to transfer heat loads generated during design basis accidents based on the design temperature of the ultimate heat sink defined for accident conditions.

4.43. The heat transfer chain should be designed in accordance with the recommendations provided in para. 3.58.

4.44. The heat transfer chain should be designed to have capabilities to fulfil simultaneously the following functions in the event of a design basis accident:

(a)  Transferring residual heat from the reactor coolant system to the ultimate heat sink;
(b)  Transferring heat from the spent fuel pool cooling system to the ultimate heat sink;
(c)  Transferring heat from the containment to the ultimate heat sink;
(d)  Transferring heat from water cooled components.

**Residual heat transfer for design extension conditions**

4.45. Conditions necessitating additional equipment (i.e. safety features for design extension conditions) are dependent on the technology and design of the reactor, and they should be postulated by applying a deterministic approach in combination with a Level 1 probabilistic safety assessment. In particular, attention should be given to the following:

(a)  Residual heat transfer to the ultimate heat sink should be possible in the event of station blackout. This can be achieved, for example, by the cooling chain being supplied by the alternate AC power source and/or the passive secondary residual heat removal system.

(b)  The need to transfer residual heat to the ultimate heat sink in the event of a loss of the cooling chain designed for design basis accidents should be evaluated. This can be achieved, for example, by means of the passive secondary residual heat removal system or by credit of the heat transfer chain for design extension conditions with significant fuel damage.

4.46. Additional safety features for design extension conditions should be implemented in accordance with the recommendations provided in paras 3.38, 3.40, 3.54 and 3.58.

# 5. SPECIFIC CONSIDERATIONS IN DESIGN OF THE REACTOR COOLANT SYSTEM

5.1.  The reactor coolant system forms a pressure retaining boundary for the reactor coolant and is therefore a barrier to radioactive releases, which is to be preserved to the extent possible in all operational states and accident conditions. The reactor coolant system transports the coolant, and thereby heat, from the reactor core to the steam generators (for PWRs and PHWRs) or directly to the turbine generator (for BWRs). The reactor coolant system also forms part of the route for the transfer of heat from the reactor core to the ultimate heat sink during shutdown and in all transient conditions that are considered in the design of the reactor coolant system. The reactor coolant system includes the reactor pressure vessel, the piping and pumps for the circulation of the coolant, and (for PWRs and PHWRs) the steam generators and pressurizer.

5.2. For PHWR technology:

(a)  The key process systems for a PHWR comprise the primary heat transport system, including the shutdown cooling system and the moderator system. The primary heat transport system circulates pressurized heavy water through the fuel channels to remove the heat produced in the fuel channels. This heat is transferred to ordinary light water in the steam generators located inside the reactor building. During shutdown periods, the shutdown cooling system is used in conjunction with the primary heat transport system to remove residual heat from the fuel.
(b)  The reactor coolant system comprises the primary coolant pumps, the primary side of the steam generators, the reactor inlet and outlet headers, the fuel channels, the pressurizer, the feeders and the piping up to and including the isolation devices, and the shutdown cooling system, which comprises pumps and heat exchangers.
(c)  The heavy water moderator is circulated through the calandria and cooled in a relatively low temperature, low pressure system. The system comprises pumps and heat exchangers. The heat exchangers remove the heat generated in the moderator and the heat transferred to the moderator from the fuel channels. Helium is used as a cover gas over the heavy water moderator in the calandria.


STRUCTURAL DESIGN

5.3.  Paragraphs 5.4–5.16 provide recommendations on meeting Requirement 47 of SSR-2/1 (Rev. 1) [1].


5.4.  Technical specifications should be established for the design and manufacture of the reactor coolant pressure boundary, and of the secondary side pressure boundary for PWRs and PHWRs, in order to achieve a high reliability of these components. These specifications should be established in accordance with the latest edition of proven codes and standards, taking into account regulatory requirements and available experience, and should include specifications for the following:

(a)  Analysis of the relevant potential damage modes and the selection of appropriate materials with proven structural characteristics;
(b)  Comprehensive identification of loads and load combinations and appropriate margins with regard to the failure criteria;

(c)    Manufacture and inspection on the basis of proven and qualified industrial practices;

(d)    An in-service inspection programme to verify that the original quality of equipment is maintained during its lifetime, in particular that cracks or defects of significance to safety do not occur.

5.5.    A high degree of confidence in the design and manufacture of the large components of the reactor coolant system should be provided in order that the failure of such components does not need to be regarded as a postulated initiating event in the plant design (i.e. because the consequences of such a failure cannot be reasonably mitigated).

5.6.    The following types of failure mode should be considered in the design, in accordance with the criteria specified in relevant codes:

(a)    Excessive plastic deformation;
(b)    Elastic or elastoplastic instability (buckling);
(c)    Progressive deformation and ratcheting;
(d)    Progressive cracking due to mechanical and thermal fatigue;
(e)    Fast fracture, including brittle fracture, in the case of existing defects in the structure.

5.7.    To preserve the integrity of the reactor coolant system, any condition (in particular corrosion, stratification or ageing) that would affect the geometry or structural characteristics of equipment, or cause defects, should be identified and prevented by design, manufacture and/or operating provisions, and by in-service inspection provisions.

5.8.    The equipment of the reactor coolant system should be designed so that the stresses imposed upon it remain below the values defined for structural materials, so as to prevent a fast growth crack during normal operation, anticipated operational occurrences, design basis accidents and design extension conditions without significant fuel degradation.

5.9.    The cyclic plant conditions that might cause the appearance of cracks due to fatigue should be identified for each reactor coolant system component. These plant conditions should be identified at the design stage in order to be monitored during the plant operation, and an admissible frequency of occurrence should be assigned to each condition according to the usage factor assessment of each component.

5.10. Adequate systems with appropriate accuracy, reliability and response time should be installed to detect a coolant leak and to quantify any such leak in operational states.


DESIGN BASIS LOADS AND LOAD COMBINATIONS

5.11. The structural design of the reactor coolant pressure boundary, and also the secondary side pressure boundary for PWRs and PHWRs, should be established on the basis of a limited number of loads and load combinations that define the design envelope of loads to which the equipment could be subjected over its lifetime. The design should take into account the normal operation of the plant, anticipated operational occurrences and accident conditions caused by the postulated initiating events, and the site hazards considered in the design basis of the equipment.

5.12. At low operating temperatures, the ductility and fracture resistance of some materials might be significantly lower than at normal operating temperatures. Where such materials are used for the manufacture of pressure retaining components, the allowable loads at low operating temperatures should be defined, the permitted operational ranges for pressure and temperature should be determined and a protection system (e.g. the overpressure protection system) should be implemented to prevent brittle fracture of the material of the component, taking into account the specified ranges of pressure and temperature established for the normal operation of the plant.

5.13. Stresses caused by normal service conditions and upset conditions (see para. 3.79) should be less than the stress limits specified for these categories of loading conditions. The design temperature should not be exceeded, and it is good practice not to exceed the design pressure. The cumulative usage factor should be less than one for each component.

5.14. For loading conditions assigned to the emergency conditions category, the design criteria should aim at preventing the fast fracture of the equipment and at avoiding excessive deformation or buckling. Stresses should be less than the stress limits specified for this category of loading conditions. The pressure reached during emergency conditions may be allowed to exceed the design pressure, provided that the excess is limited in magnitude and time (e.g. it should not exceed 110% of the design pressure).

5.15. For loading conditions assigned to the faulted conditions category, the design criteria should aim to preserve the integrity of the equipment. Stresses should be less than the stress limits specified for this category of loading conditions (e.g. it should not exceed 130% of the design pressure).

5.16. The thermohydraulic conditions in the reactor coolant system should be monitored throughout the lifetime of the plant in order to identify and record situations that might cause fatigue to reactor coolant system equipment. This monitoring should prove that the frequency of occurrences assigned for each plant condition (see para. 5.9) is not exceeded throughout the lifetime of the plant, and that there is a minimal risk of cracking induced by fatigue.


CONTROL OF COOLING CONDITIONS IN OPERATIONAL STATES

5.17. Paragraphs 5.18 and 5.19 provide recommendations on meeting Requirement 49 of SSR-2/1 (Rev. 1) [1].

5.18. Design provisions should be implemented for the monitoring, display and control of important reactor coolant system parameters (e.g. reactor coolant system pressure and temperature, reactor coolant system water inventory, steam and feedwater flow (for BWRs), steam generator pressure and water levels (for PWRs and PHWRs)) to maintain these parameters within the ranges specified for normal operation and anticipated operational occurrences, and to detect any early deviation from the normal values. Maintaining these parameters within the ranges for normal operation contributes to ensuring adequate cooling conditions for the fuel.

5.19. Structures, systems and components that maintain adequate cooling conditions for the fuel should be classified as items important to safety and should be designed and manufactured accordingly (see paras 3.62–3.66).


PRESSURE CONTROL AND OVERPRESSURE PROTECTION

5.20. The recommendations in paras 5.21–5.27 apply to the design of the pressure control of the reactor coolant system and the secondary side for PWRs and PHWRs.

5.21. The concept of defence in depth should be applied in the design of the pressure control of the reactor coolant system and the secondary side. According to this concept, systems and components with different capacities should be used for pressure control to ensure that the preventive measures are proportional to the severity of anticipated operational occurrences or accident conditions.

5.22. In the design, the diversity principle should be applied between the pressure control system and the overpressure protection system to reduce the likelihood of common cause failures.

5.23. If the pressurizer can be isolated from the reactor coolant system in certain operating conditions (e.g. during heat-up or cooldown of a PHWR), the pressure and inventory control system should include alternative means of controlling the pressure and inventory in the reactor coolant system, such as a set of automatically controlled feed and bleed valves. In such cases, the pressurizer should have independent safety devices and/or relief devices.

5.24. Systems intended for the control of pressure should be designed to prevent, in normal operation and anticipated operational occurrences, the operation of the safety valves (or, for BWRs, the safety relief valves). The release of primary coolant should be minimized, and it should not be discharged into the containment.

5.25. The settings and performance of systems designed to control operating conditions should be determined on the basis of a realistic response of the plant to such conditions.

5.26. The pressure control system of the reactor coolant system should be designed to maintain the pressure within the limits that are set to ensure the cooling of the fuel in operational states (as long as two-phase conditions are maintained in the pressurizer).

5.27. Pressure control in the reactor coolant system and the secondary side circuit should be ensured even in the event of a loss of off-site power.

5.28. Paragraphs 5.29–5.41 provide recommendations on meeting Requirement 48 of SSR-2/1 (Rev. 1) [1] for PWRs, BWRs and PHWRs and also apply to the overpressure protection system of the secondary side for PWRs and PHWRs.

5.29. The overpressure protection devices should include redundant safety valves (or, for BWRs, safety relief valves). The setting of the safety valves should be such that safety valves (or, for BWRs, safety relief valves) open in sequence for different levels of pressure to avoid unnecessary discharge of coolant.

5.30. An overpressure protection system should be implemented to preserve the structural integrity of the reactor coolant pressure boundary by keeping, in conjunction with the reactor shutdown system, the pressure below the design limits specified for the different categories of postulated initiating event.

5.31. For PWRs and PHWRs, an overpressure protection system should be implemented to preserve the structural integrity of the secondary side pressure boundary by keeping, in conjunction with the reactor shutdown system, the pressure below the design limits specified for the different categories of postulated initiating event.

5.32. The discharge capacity of the overpressure protection system should be designed to meet the pressure limits prescribed by proven industry codes and should apply the design rules specified by these codes. The typical approach includes the following:

(a) Analyses do not credit systems that are not safety classified unless the operation of such systems can aggravate the consequences of the initiating event.
(b) With regard to the criteria to be met, safety classified systems are assumed to operate at a less favourable performance.
(c) The discharge capacity of the safety valves is determined on the basis of the applicable design standard.
(d) The total discharge capacity credited in the analysis is calculated taking into account the sequential opening of the safety valves and that at least one safety valve fails to open (or more safety valves fail, for systems with more safety valves). This independent failure need not be considered in the analysis of overpressure transients initiated by multiple failures.
(e) Loss of off-site power is combined if it can aggravate the consequences of the initiating event.

5.33. The overpressure protection devices should be designed to keep water hammer effects as low as possible.

5.34. Equipment ensuring the integrity of the reactor coolant pressure boundary and/or the integrity of the secondary side pressure boundary should be supplied by the uninterruptible power source.

5.35. Shut-off valves should not be placed in the discharge line of a safety valve (or, for BWRs, a safety relief valve), or between a safety valve (or, for BWRs, a safety relief valve) and the item being protected. When a relief valve is used for pressure control, its reliable closing should be secured by means of a shut-off valve on the relief line.

5.36. Safety valves, safety relief valves and relief valves should be provided with a position indicator that is independent of the control equipment.

5.37. The outlet of the steam relief valves should be monitored in order to detect leakages.

5.38. Due consideration should be given to the layout of the safety valves, their pilot valves and connecting piping to prevent the accumulation of non-condensable gases and condensate, and thereby avoid the adverse effects of such accumulation.

5.39. The valves used for overpressure protection and the associated piping should be designed to discharge steam, steam–water mixtures and water.

5.40. The spurious opening of a safety valve (or, for BWRs, a safety relief valve) should be prevented, and the frequency of such spurious opening should not be higher than the frequency considered for loss of coolant accidents.

5.41. Components that can increase pressure in the primary circuit (e.g. pressurizer heaters or make-up pumps for PWRs and PHWRs) should be equipped with a system that stops the operation of the component to prevent an inadvertent pressure increase.

5.42. Paragraph 5.43 provides recommendations on meeting Requirement 20 of SSR-2/1 (Rev. 1) [1].

5.43. The reactor coolant system overpressure protection system should also be designed to preserve the integrity of the reactor coolant pressure boundary in the event of postulated sequences involving multiple failures. Typically, the design of the overpressure protection system should be adequate to limit the pressure in the event of an anticipated transient without scram (if relevant).

5.44. The reactor coolant pressure boundary comprises pressure retaining components of the reactor coolant system that cannot be isolated from the reactor. The reactor coolant pressure boundary includes the following valves:

(a)   The outermost containment isolation valve in the reactor coolant system piping that penetrates the primary reactor containment (if relevant);
(b)   The reactor coolant system safety valves (or, for BWRs, safety relief valves);
(c)   The second of two isolation valves for piping connected to the reactor coolant system and whose failure would result in a leakage that cannot be compensated for by the normal water make-up system;
(d)   The first isolation valve for piping connected to the reactor coolant system and whose failure would result in a leakage that can be compensated for by the normal water make-up system (if relevant);
(e)   The second isolation valve as seen from the reactor side in the main steam piping and feedwater piping (for BWRs).

5.45. Paragraphs 5.46–5.53 provide recommendations on meeting para. 6.13 of SSR-2/1 (Rev. 1) [1].

5.46. Isolation devices between the reactor coolant pressure boundary and connected piping or components whose failure would result in a leakage that cannot be compensated for by the normal water make-up system should be designed to close quickly and reliably in order to limit the loss of coolant. The loss of coolant caused by such a failure should not necessitate the operation of the emergency core cooling system.

5.47. Consideration should be given to the characteristics and importance of the isolation device and the necessary reliability. Isolation devices should either be closed (the usual position) or should close automatically when necessary without the need for off-site electrical power. The response time and closure time should be in accordance with the acceptance criteria defined for postulated initiating events. In particular, to keep the full efficiency of the emergency core cooling system, all the systems connected to the reactor coolant pressure boundary should be automatically isolated in a timely manner, unless these systems are necessary to meet the criteria applicable to a loss of coolant accident.

5.48. The isolation of the reactor coolant pressure boundary should be designed in accordance with the single failure criterion if the loss of coolant resulting from a break of the connected piping cannot be compensated for by the normal water make-up system.

5.49. The isolation of the reactor coolant pressure boundary should be ensured even in the event of a loss of off-site power, and should preferably be supplied by the uninterruptible power source.

5.50. The isolation valves for the reactor coolant pressure boundary should be designed to return to a safe position in the event of a loss of power or a loss of the compressed air supply.

5.51. Adequate means should be provided in connected systems that operate at lower pressures to prevent the overpressurization of such systems and possible loss of coolant accidents outside the containment.

5.52. Isolation devices should be designed and manufactured in compliance with the design requirements that apply to reactor coolant system components.

5.53. Provisions for testing the leaktightness of the isolation valves for the reactor coolant pressure boundary should be implemented.


POSTULATED INITIATING EVENTS

5.54. The failure of any structure, system or component — and operator errors whose consequences would modify the reactor coolant system conditions or loads defined for normal operation — should be identified and categorized on the basis of their estimated frequency of occurrence. Typical examples include:

(a)     Loss of off-site power.
(b)     Malfunctioning of pressure control systems:
            — Reactor pressure vessel water level;
            — Reactor coolant system recirculation flow;
            — Feedwater heating (for BWRs);
            — Pressurizer and steam generator level (for PWRs and PHWRs).
(c)     Loss of the main condenser vacuum.
(d)     Piping breaks.
(e)     Spurious opening of a safety valve (or, for BWRs, a safety relief valve).
(f)     Loss of forced coolant circulation.
(g)     Reactor coolant pump failure.
(h)     Positive core reactivity insertion.

INTERNAL HAZARDS

5.55. The layout of reactor coolant system piping supplemented by local protection devices (e.g. anti-whipping devices and shields) should be such that 'domino effects' are prevented in the event of a high energy pipe break. Typical examples include:

(a)   A break of a reactor coolant leg should neither propagate to a neighbouring reactor coolant system leg nor to the main steam piping or feedwater piping (for PWRs and PHWRs).
(b)   A break of the main steam piping or feedwater piping should neither propagate to neighbouring main steam or feedwater piping nor to reactor coolant loops.
(c)   A break of pressurizer piping should not propagate to neighbouring pressurizer piping (for PWRs and PHWRs).

5.56. Fail-safe instrumentation and layout provisions should be implemented to protect the instrumentation and to ensure the actuation of necessary automatic actions during accident management.


EXTERNAL HAZARDS

5.57. The integrity of the reactor coolant pressure boundary should be maintained in the event of SL-2 seismic loads.

5.58. Reactor coolant system components that are not part of the reactor coolant pressure boundary and that are not seismically classified should be reliably isolated from the reactor coolant pressure boundary by isolation devices that are seismically classified and qualified to operate under SL-2 seismic loads.

5.59. Reactor coolant system components that are necessary to put the reactor into safe shutdown conditions should be seismically classified to keep their integrity and qualified to operate under SL-2 seismic loads.

5.60. Reactor coolant system components that are necessary to put the reactor into safe shutdown conditions should be designed to keep their integrity and to operate in the event of external hazards causing a high energy impact on the containment.

LAYOUT

5.61. The layout and location of the piping and equipment should facilitate the establishment of natural circulation, allowing the removal of core decay heat.

5.62. The layout and location of the piping and equipment should be such that flow induced vibration, flow accelerated corrosion, ageing effects, acoustic excitation, thermal fatigue and the accumulation of radioactive material are all minimized. The layout of the piping should also be such as to prevent the accumulation of combustible gases.

5.63. The layout and location of the piping and equipment should provide sufficient accessibility to allow periodic testing, maintenance and inspection to be conducted, including maintenance and inspection of welds and piping supports.


DESIGN LIMITS

5.64. Design limits that are not to be exceeded in each plant state should be defined for reactor coolant system components. Parameters for which design limits should be defined include the following:

(a)  Pressure and temperature;
(b)  Maximum cooling rate and maximum heating rate for normal operation;
(c)  Maximum differential temperature ($\Delta T_{max}$) between the hot leg and the pressurizer (for PWRs);
(d)  Maximum differential primary and secondary pressures ($\Delta P_{max}$) (for PWRs);
(e)  Maximum reactor coolant system leak rate;
(f)  Maximum leak rate from reactor coolant system to steam generator (for PWRs and PHWRs);
(g)  Limits with regard to the brittle fracture of the reactor pressure vessel (for PWRs);
(h)  Individual component parameters (e.g. $\Delta P$ for reactor coolant pump seals and T seals).

The recommendations provided in NS-G-2.2 [16] should also be considered, where relevant.

SAFETY CLASSIFICATION

5.65. Components that are part of the reactor coolant pressure boundary should be safety classified so that they are designed and manufactured in compliance with the highest standards used by the industry for nuclear application (e.g. Refs [24, 25] and similar standards).

5.66. Other components should be classified with due account taken of the consequences of failure, in terms of both the fulfilment of the safety function and the prevention of a radioactive release, in accordance with the recommendations provided in SSG-30 [15] (see Ref. [26] for further guidance).


ENVIRONMENTAL QUALIFICATION

5.67. Reactor coolant system components should be designed and qualified for the harshest environmental conditions that could prevail inside the containment prior to or during accident conditions without significant fuel degradation until their service life is completed. This typically applies to the following:

(a)     Reactor coolant pressure boundary components (for integrity);
(b)     Reactor coolant pressure boundary isolation devices (for operability);
(c)     Overpressure protection components (for operability);
(d)     Reactor coolant system components designed to protect the reactor pressure vessel against brittle fracture (for operability);
(e)     Reactor coolant system components necessary to put the reactor into safe shutdown conditions (for operability);
(f)     Reactor coolant system components designed to depressurize the reactor coolant system, allowing the use of low pressure pumps in accident conditions;
(g)     Reactor coolant system components designed to depressurize the reactor coolant system to prevent direct containment heating loads caused by failure of the reactor pressure vessel at high pressure (for operability).

5.68. The selection of materials to be used for gaskets and seals should be based on their suitability to maintain their capability for all operational states and accident conditions.

PRESSURE TESTS

5.69. The hydrostatic pressure test of the reactor coolant system should be performed at the commissioning stage and should be repeated periodically, possibly with different criteria. The aim of these tests should be the following:

(a) To prove the absence of permanent deformation when the structure is brought to a pressure close to the elastic limit;
(b) To check for leaks that were not detected previously.

The pressure test operating mode and pressure level are usually defined in national regulations and are reflected in industry codes and standards.

5.70. The reactor coolant system equipment should not experience any damage during the pressure test.


VENTING

**For pressurized water reactors**

5.71. In order to prevent disruption to the natural circulation of the reactor coolant, remotely operated valves should be provided to vent non-condensable gases in accident conditions.

5.72. The reactor coolant system venting should be possible in the event of a loss of off-site power.

5.73. The capability for venting should be consistent with the capacity of the make-up system.

**For boiling water reactors**

5.74. In order to accommodate water level changes in the reactor pressure vessel during shutdown and startup, valves that can be operated remotely should be provided to vent the reactor pressure vessel head.

5.75. During normal operation, venting of the reactor pressure vessel head and piping should be possible to prevent the accumulation of non-condensable gases.

5.76. Venting should be effective despite a single failure.

5.77. The reactor coolant system components necessary for the venting should be qualified for accident conditions relevant for their intended use.

5.78. Control of the venting operation should be possible from the main control room.

**For pressurized heavy water reactors**

5.79. Provisions should be implemented to limit the concentration of non-condensable gases in the primary coolant in normal operation.

SPECIFIC DESIGN ASPECTS

**Reactor pressure vessel**

5.80. The design considerations for the pressure vessel should include the following:

(a) The number of welds in the pressure vessel should be minimized; in particular, the need for welds in the active core region should be assessed.
(b) Pressure limits as well as allowable heating and cooling rates as a function of temperature should be established for the pressure vessel. Changes in the brittle–ductile transition temperature of the vessel material adjacent to the core (the 'beltline region') due to neutron irradiation and thermal embrittlement should be accounted for.
(c) The vessel should be designed to withstand all the cyclic loads that are expected to occur throughout the lifetime of the plant. The design documentation should include clear specifications of those loads that are necessary for the determination of the cumulative usage factor.
(d) The choice of material, the structural design, the welding and the heat treatment should be such as to ensure a sufficiently ductile state of the pressure vessel material throughout the lifetime of the plant. The ductility of the pressure vessel wall facing the core should be ensured by limiting the maximum neutron flux and by using base metal and weld metal of a chemical composition such as to keep radiation embrittlement at an acceptable level.
(e) The design of the pressure vessel should be such that it can withstand pressurized thermal shocks without losing its integrity.
(f) Thermal cycling at the vessel nozzles and penetrations should be minimized, including the use of thermal sleeves, as appropriate.
(g) A corrosion resistant cladding should be applied to the interior of the reactor pressure vessel.

5.81. If advanced materials are to be used for the reactor pressure vessel, samples of these materials should be subjected to a fast neutron flux with a high lead factor compared to the vessel wall, and exposed to the environmental conditions of the pressure vessel. The samples should be examined periodically throughout the lifetime of the plant to monitor changes in mechanical properties (in particular, ductility and toughness) and to enable predictions to be made of the behaviour of the material in sufficient time to allow for corrective measures, if necessary.

5.82. For a design with an in-vessel retention strategy, the robustness of the reactor pressure vessel to sustained loads caused by such severe conditions should be demonstrated with a high degree of confidence.

**Reactor pressure vessel internals (for PWRs and BWRs)**

5.83. Pressure vessel internals should be removable to facilitate maintenance, replacement and in-service inspection. The use of bolted connections should be considered instead of welded connections, where appropriate.

5.84. Pressure vessel internals should be designed to withstand loads associated with operational states and accident conditions without significant fuel degradation and should maintain the capabilities to support the core and permit the cooling of the fuel elements, and to insert the control rods into the core to shut down the reactor.

5.85. Pressure vessel internals should be designed to do the following:

(a)   Properly channel the coolant flow through the vessel and the fuel;
(b)   Prevent unacceptable flow induced vibration;
(c)   Minimize susceptibility to stress corrosion cracking;
(d)   Accommodate asymmetric blowdown loads caused by pipe ruptures;
(e)   Ensure that fuel design limits are not exceeded in normal operation or anticipated operational occurrences.

**Fuel channel assemblies (for PHWRs)**

5.86. The fuel channels should be designed to provide a pressure boundary with low neutron absorption to support and locate the fuel bundles. The fuel channels should allow for controlled flow of the coolant around and through the fuel bundles.

5.87. The fuel channel assemblies should be designed to meet all applicable requirements for their specified design life.

5.88. The fuel channel design should permit continuous gas flow in the annulus between the pressure tube and the calandria tube to allow leak before break detection.

5.89. All materials used in the fuel channel assembly should be capable of withstanding prolonged exposure to radiation, high purity heavy water and the annulus gas (the gas between the pressure tubes and the calandria tubes).

5.90. The design conditions for the fuel channels should be taken as the most adverse combination of temperature and pressure anywhere along the length of the pressure tube.

5.91. The fuel channels should be designed and manufactured in accordance with proven codes and standards, taking into consideration available experience, including operating experience.

5.92. Prototype rolled steel joints employed in the reactor coolant system should be tested for pullout strength. The axial pullout load should be at least three times the design condition total axial load when the test is performed at design temperature.

5.93. Fuel channels should be designed to withstand all the cyclic loads that are expected to occur throughout the lifetime of the plant. The design documentation should include clear specifications of those loads that are necessary for the determination of the cumulative usage factor.

5.94. Welds that cannot be inspected in-service should be limited to the extent possible, and analyses should be performed to assess the consequences of the failure of such welds.

5.95. The design should provide a means for allowing the reliable detection of defects in the fuel in the core during normal operation.

**Reactor coolant pumps (for PWRs and PHWRs) and reactor recirculation pumps (for BWRs)**

5.96. The design of reactor coolant pumps should be such that the following safety parameters are adequate:

(a)  Pump performance characteristics, including head–flow characteristics, flow coastdown rate, and single phase and two-phase pump performance;
(b)  Pump operating parameters (e.g. speed, flow and head);

(c)     The net positive suction head that is necessary to avoid cavitation;

(d)     Pump seal design and performance (including seal temperature limitations, if applicable);

(e)     Provisions for vibration monitoring.


5.97. The design of the reactor coolant system pumps should be such as to provide an adequate flow of coolant with suitable hydraulic parameters to ensure that the fuel design limits are not exceeded in operational states.


5.98. The design of the reactor coolant system pumps should be such as to provide an adequate flow of coolant with suitable hydraulic parameters to ensure that the structural limits for the fuel and for the reactor coolant system equipment (including the reactor pressure vessel internals) are not exceeded in operational states and in accident conditions without significant fuel degradation.


5.99. The reactor coolant system pumps should have adequate flow coastdown characteristics in the event of a pump trip in anticipated operational occurrences and in accident conditions without significant fuel degradation to avoid undesirable thermohydraulic conditions of the reactor coolant with regard to the integrity of the fuel.


5.100.   The design of reactor coolant system pumps should be such that neither adverse thermohydraulic conditions in the reactor coolant system nor pump malfunctions result in the generation of missiles. Alternatively, provision should be made to protect items important to safety from any such missiles.


5.101.   The correct operation of pads and bearings should be monitored, and an automatic trip of the reactor coolant system pumps should be implemented in order to prevent operation under excessive vibration that could result in shaft failure.


5.102.   Seal leakage in reactor coolant pumps should be controlled by maintaining adequate cooling of seal systems in any plant states without significant fuel degradation. In normal operation, seal leakage should be compensated for; in plant states where compensation is not available, seal leakage should preferably be isolated. The reactor coolant pump should be automatically tripped in the event that seal operational parameters cannot be maintained, in order to prevent any further damage to the seal system.

**Relief and safety valves (for PWRs and PHWRs) or safety relief valves (for BWRs)**

5.103.  If compressed air is needed to operate relief valves, dedicated pneumatic accumulators should be implemented for the valves to ensure a minimum and specified number of openings and closures. The number of times that each valve has to be capable of being opened utilizing the pneumatic accumulators without recharging should be defined.

5.104.  When a relief valve is used for pressure control, its reliable closing should be secured by means of a shut-off valve on the relief line.

5.105.  Where the overpressure protection is ensured by pilot operated safety valves (BWRs and PWRs), shut-off valves should not be installed in the pilot line for opening the safety valve. If an exception is made to facilitate testing or maintenance, or to prevent a safety valve from being stuck open, the inadvertent closing of the shut-off valve should be reliably prevented.

**Steam generators (for PWRs and PHWRs)**

5.106.  Steam generator tubes are part of the reactor coolant pressure boundary and hence should be designed in accordance with the recommendations in paras 3.44, 3.107–3.115 and 5.4–5.16.

5.107.  The steam generator tubes and the internal structures of the steam generators should be designed for the maximum stresses and most severe fatigue conditions expected to occur in operational states and in accident conditions without significant fuel degradation (e.g. should be designed to withstand loads from a loss of coolant accident and a main steam line break).

5.108.  The flow pattern in the steam generators should be optimized to prevent the occurrence of areas of stagnant flow (to avoid the accumulation of precipitates) and unacceptable flow induced vibration of the tubes.

5.109.  The design of the steam generators should provide an adequate system for tube leak detection and alarms.

5.110.  Overfilling of the steam generators should be prevented by design provisions.

5.111.    Loads such as those due to water hammer, overfilling and thermal and/or hydraulic stratification should be addressed for the operating modes in which they might occur.

5.112.    The design should include blowdown provisions to control and remove the solids (sludge) that could accumulate in areas of stagnant flow.

5.113.    The design should include provisions for sampling water and steam from relevant locations in the secondary side.

5.114.    The design should allow for inspection of the steam generator tubes and the primary and secondary steam separators over their entire length. The equipment and procedures for examination of the tubes should be capable of detecting and locating significant defects or degradation.

5.115.    The design should also provide for the following:

(a)    Control of the pH and oxygen concentration;
(b)    Limitation of the concentration of contaminants and impurities in the secondary side of the steam generators;
(c)    Addition of chemical additives to the feedwater;
(d)    Monitoring of the conductivity of the sampled fluid and monitoring for contamination of the sampled fluid.

5.116.    The allowance for fouling, tube plugging and for maximum allowable tube leakage should be specified.

5.117.    Design provisions to prevent tube fretting should be implemented.

5.118.    Design provisions to perform sampling in the steam generator should be implemented.

5.119.    Steam generator tube material, tube plate material and welding materials should be able to withstand the corrosion and erosion impact of the primary or secondary coolant (as appropriate), including undersludge corrosion.

5.120.    Electrochemical interactions between steam generator tube materials and tube support materials should be prevented.

**Piping system**

5.121.   Piping should be arranged to limit the possibility of the accumulation of non-condensable gases.

5.122.   Capability for venting and draining the piping system should be provided.

5.123.   The design of piping supports should be in accordance with the applicable piping system design standard. The stress assessment for piping and components should be conducted in compliance with applicable codes and standards.

5.124.   Flow restrictors should be included in the main steam lines for BWRs (or the steam generator outlet for PWRs and PHWRs) to limit the rate of loss of coolant following a main steam line break inside or outside the containment. For BWRs, it should be ensured that the core remains fully covered with water before closing the main steam isolation valves.

**Leak before break or break preclusion piping**

5.125.   If the leak before break or break preclusion concept is applied for piping, the specific requirements to be met for design, manufacturing and operation should be defined.

5.126.   Leak detection systems should perform in a manner that is consistent with leak before break assumptions.

5.127.   Regardless of the very low probability of piping failure, the consequences of a double ended break of a pipe should be analysed using appropriate rules with regard to:

(a)   Core cooling capacity;
(b)   Pressure buildup inside the primary containment;
(c)   Environmental qualification of equipment.

**Leak detection system**

5.128.   Provisions should be made for the detection of any leakage of reactor coolant and, to the extent practicable, for the identification of the location of the leak. Provisions should also be made to quantify and collect fluid that has leaked from the reactor coolant system. These provisions should be adequately complemented by indicators and alarms in the main control room.

**Insulation materials**

5.129.   The insulation material used inside the primary containment should be selected so as to prevent clogging of the sump strainers and filters in the event of high energy pipe breaks. Reflective metallic insulation should be used to the extent that is reasonably achievable.

5.130.   For a design relying on an in-vessel retention strategy and ex-vessel cooling, the following design considerations apply to the reactor vessel insulation:

(a)   A means should be provided to allow water free access to the region between the reactor vessel and insulation, and the design of the associated water inlets should minimize the pressure drop during ex-vessel cooling to permit water inflow to cool the vessel.
(b)   A means should be provided to allow the steam generated from water in contact with the reactor vessel to escape from the region surrounding the reactor vessel.
(c)   The insulation support frame and the insulation panels should form a structurally reliable flow path for water and steam.


# 6. SPECIFIC CONSIDERATIONS IN DESIGN OF THE ASSOCIATED SYSTEMS FOR PWR TECHNOLOGY

## SYSTEMS FOR CONTROL OF COOLANT INVENTORY AND CORE REACTIVITY IN OPERATIONAL STATES

6.1.   The control of the reactor coolant system water inventory in normal operation is performed by the chemical and volume control system. The chemical and volume control system is also designed to control the reactor coolant system pressure when the reactor coolant system pumps are shut down by spraying the reactor coolant system pressurizer and to adjust the boric acid concentration of the primary coolant in power operation and shutdown modes. Typical functions performed by the chemical and volume control system are:

(a)   Control of the reactor coolant inventory;
(b)   Control of the reactor coolant system pressure in shutdown modes when the reactor coolant pumps are unavailable;
(c)   Control of the core reactivity;

(d)    Supply of seal water to the reactor coolant pumps;

(e)    Control of the chemistry of the reactor coolant;

(f)    Cleanup and purification of the reactor coolant.

6.2.    These functions are mainly dedicated to normal operation and are not usually performed during accidents. However, parts of the system might be used to reach a safe shutdown state following anticipated operational occurrences or accidents.

**Control of the coolant inventory**

6.3.    The chemical and volume control system should be designed to provide a let-down function for reactor coolant expansion during plant heat-up and to provide the make-up for reactor coolant shrinkage during plant cooldown at heat-up and cooldown rates.

6.4.    The chemical and volume control system should be designed to provide make-up water in the event of power decreases and to provide a let-down function in the event of power increases.

6.5.    For any mode of normal operation or event that does not require the operation of safety systems, the chemical and volume control system should provide and maintain sufficient reactor coolant inventory to ensure core cooling such that the fuel design limits are not exceeded, and provide sufficient flow to the reactor coolant pump seals for the pressure boundary integrity to be maintained.

**Control of core reactivity**

6.6.    The chemical and volume control system should be designed to adjust the boric acid concentration in the reactor coolant system in order to control the axial offset of the core during power operation.

6.7.    The chemical and volume control system should have capabilities to achieve the necessary boric acid concentration in the reactor coolant system for refuelling operations.

6.8.    The chemical and volume control system should have capabilities to achieve the necessary boric acid concentration in the reactor coolant system for power operation for conditions during the whole refuelling cycle.

6.9.    The chemical and volume control system should have capabilities to prevent or limit uncontrolled dilution of the reactor coolant system.

# SYSTEMS FOR HEAT REMOVAL IN OPERATIONAL STATES

## Heat removal in power operation and hot shutdown modes

6.10. The heat, including residual heat, generated in the core in power operation and hot shutdown modes is transferred from the reactor coolant system to the steam generators. The heat removal function is ensured by the main feedwater system and the main steam system. The main feedwater system comprises the main feed pumps, control valves and isolation valves. In some designs, there are pumps dedicated to low power and shutdown modes, and they belong to a system designated as the 'startup and shutdown feedwater system'. The main steam system includes the main steam lines, isolation valves, safety valves and dump valves to the main condenser.

6.11. The main feedwater and main steam systems are primarily designed to remove heat generated by the reactor at full power but should also have capabilities to remove and transfer residual heat to an ultimate heat sink after the reactor is shut down.

6.12. Residual heat removal capabilities should be designed to cool down the reactor coolant system from hot shutdown conditions to a primary pressure and temperature compatible with the operation of the residual heat removal system.

6.13. The main feedwater system should have capabilities to feed steam generators at rated temperatures and to control the steam generator levels within the range specified for operational states.

6.14. The failure of one feedwater pump should not lead to a reactor trip.

6.15. The main feedwater injection should be automatically stopped after a reactor trip in order to prevent excessive cooling of the core.

6.16. The overfilling of steam generators should be reliably prevented.

6.17. In the event of an uncontrolled and excessive steam generator depressurization (e.g. in the event of a main steam pipe or main feedwater pipe break), the affected steam generator should be reliably isolated from other steam generators.

6.18. Each steam generator should be able to be independently and reliably isolated.

6.19. Adequate monitoring for radioactivity should be available to allow detection of a leaking or ruptured steam generator tube. The accuracy of this monitoring should be adequate to meet the limits specified for radiological consequences in design basis accidents.

6.20. The leaktightness of steam generator isolation valves should be adequate to meet the limits specified for radiological consequences in the event of a steam generator tube rupture.

6.21. The main steam system should provide capability to automatically and manually bypass the turbine and discharge steam directly to the condenser. The capacity of the bypass should be adequate to accommodate a full load rejection.

6.22. The main steam system should be designed such that one main steam line break could not lead to the depressurization of more than one steam generator despite a single failure in the isolation of the main steam piping.

6.23. The main steam and feedwater piping should be routed, protected and restrained to prevent concurrent pipe breaks (e.g. main steam or feedwater pipe breaks, and primary pipe breaks).

6.24. The minimum heat removal capacity should be designed to remove residual heat despite a single failure postulated in any component necessary for removing residual heat.

6.25. Residual heat removal should be possible in the event of a loss of off-site power.

6.26. The systems operated to remove residual heat in a hot shutdown mode should be designed to keep their operability in the event of SL-2 seismic loads.

**Residual heat removal mode**

6.27. In the cold shutdown mode of normal operation, the residual heat is transferred from the reactor coolant system to the cooling chain by the residual heat removal system. The residual heat removal system can be connected after the reactor coolant system has been cooled down by the steam generators. A residual heat removal system train comprises a shutdown cooling pump and a heat exchanger with the intermediate cooling system; it takes suction water from the reactor coolant system and injects back into the reactor coolant system after cooling in the heat exchanger.

6.28. Paragraphs 6.29–6.40 provide recommendations on meeting Requirement 51 of SSR-2/1 (Rev. 1) [1] and supplement the generic recommendations provided in Section 3.

6.29. The design of the plant should include appropriate systems to remove residual heat from the reactor coolant system in different reactor coolant system shutdown conditions (e.g. hot shutdown conditions, cold shutdown conditions and refuelling conditions).

6.30. The heat removal capacity should be designed to cool down the reactor coolant system from hot shutdown conditions (once the reactor is shut down) to adequate conditions for refuelling operation.

6.31. The residual heat removal should be designed to control the reactor coolant system temperature and to achieve a controlled rate of cooling to the cold shutdown conditions for refuelling within an appropriate time after reactor shutdown.

6.32. The minimum heat removal capacity should be designed to remove residual heat even in the event of a single failure postulated in any component necessary for removing residual heat.

6.33. Residual heat removal should be possible in the event of a loss of off-site power.

6.34. The residual heat removal system should be designed to keep its operability in the event of SL-2 seismic loads.

6.35. Residual heat removal and transfer systems should be designed in accordance with the recommendations provided in paras 3.48, 3.54, 3.58 and 4.41–4.45 if they are also operated to remove residual heat after a design basis accident.

*Specific design aspects*

6.36. The maximum heat removal capacity should be designed taking into account operational criteria (e.g. the time delay to reach refuelling conditions) without exceeding the limits specified for the fuel and the reactor coolant pressure boundary in normal operating conditions.

6.37. The residual heat removal connection temperature should be higher than the minimum reactor coolant system temperature that can be achieved by steam generator cooling.

6.38. During power operation and hot shutdown operation, the residual heat removal system is not operated and is isolated from the reactor coolant system; there should be interlocks or other provisions to prevent the connection of the residual heat removal system to the reactor coolant system.

6.39. If part of the residual heat removal system is implemented outside the containment, adequate instrumentation should be provided to detect leakages or breaks in the system, and an adequate isolation capability should be provided to limit the radiological releases outside the containment.

6.40. Adequate instrumentation and isolation capability should be provided to detect leakages in the heat exchanger and to limit the transfer of primary water into the intermediate cooling system. These provisions should also limit the transfer of non-borated water to the reactor coolant system when it is fully depressurized.

## SYSTEMS FOR CORE COOLING AND RESIDUAL HEAT REMOVAL IN ACCIDENT CONDITIONS (EXCLUDING DESIGN EXTENSION CONDITIONS WITH CORE MELTING)

6.41. Paragraphs 6.43–6.106 provide recommendations for the design of systems necessary to cool the core, remove residual heat from the reactor coolant system and control core reactivity in all accident conditions except for design extension conditions with core melting. Recommendations for the design of the residual heat transfer chain and ultimate heat sink in accident conditions are addressed in Section 4 of this Safety Guide.

6.42. Paragraphs 6.43–6.48 provide recommendations on meeting Requirements 7, 19 and 29 of SSR-2/1 (Rev. 1) [1] and supplement the generic recommendations provided in Section 3.

6.43. The needs for different, independent and diverse systems depend on the necessary reliability of the safety systems and on potential vulnerabilities to common cause failures among their redundancies. Irrespective of these considerations, the design of the plant should be such that multiple means exist for cooling the core and removing and transferring residual heat.

6.44. Systems designed for cooling the core in design basis accidents or design extension conditions without significant fuel degradation should be independent, to the extent possible, from those designed for operating conditions and from those dedicated to core cooling in the event of design extension conditions with core melting.

6.45. The design of systems necessary for design basis accidents or for design extension conditions without significant fuel degradation should comply with the recommendations provided in paras 3.33–3.42 that are applicable to systems designed to mitigate the consequences of design basis accidents or design extension conditions without significant fuel degradation, as appropriate.

6.46. The reliability of specific safety features for design extension conditions should be adequate to meet the objective specified for the total core damage frequency.

6.47. Safety systems should be designed to meet the regulatory criteria specified for design basis accidents. The performance of these systems should be such that these criteria are met when applying the rules specified for the deterministic design basis accident analysis.

6.48. The performance of safety features for design extension conditions should be adequate to prevent accident conditions without significant fuel degradation from escalating to design extension conditions with core melting. For design, the same engineering criteria as those applied for design basis accidents can be used, but less conservative hypotheses and conditions are generally considered. However, in order to give confidence in the efficacy of the safety features for design extension conditions and to avoid cliff edge effects, key parameters should be identified and provisions should be made to address uncertainties.

**Core cooling in accident conditions**

6.49. The control of the reactor coolant system water inventory in accident conditions is performed by the emergency core cooling system in conjunction with depressurization of the reactor coolant system by the secondary side when necessary. The emergency core cooling system also performs some functions relating to the control of core reactivity. In general, the emergency core cooling system includes a combination of active and/or passive means of injection (e.g. pumps, piping and valves). The system can also include heat exchangers to remove residual heat from the containment. Recommendations concerning the sump filtration system are given in SSG-53 [22].

6.50. The main function of the emergency core cooling system is to inject borated water into the reactor coolant system in order to ensure core cooling when the reactor coolant system water inventory decreases or in the event of loss of the residual heat removal by the secondary side. The emergency core cooling system is called upon to reach a controlled state in both design basis accidents and design extension conditions without significant fuel degradation, such as:

(a) Loss of coolant accidents postulated as design basis accidents or design extension conditions;
(b) Excessive and uncontrolled reactor coolant system cooling (piping breaks on the secondary side);
(c) Steam generator tube rupture;
(d) Total loss of feedwater with availability of reactor coolant system feed and bleed.

6.51. Paragraphs 6.52–6.69 provide recommendations on meeting Requirement 52 of SSR-2/1 (Rev. 1) [1] in the event of design basis accidents and design extension conditions with a loss of primary coolant, and supplement the recommendations provided in Section 3 relevant for safety systems or for safety features for design extension conditions.

6.52. The emergency core cooling system should be designed to cool the fuel in the event of a loss of coolant accident, in accordance with the relevant design criteria for fuel, fuel cladding and core geometry.

6.53. In the event of a small pipe break, the energy removed at the break might not be sufficient for effective cooling of the fuel, and therefore complementary systems or equipment should be operated to achieve the appropriate cooling capacity (e.g. complementary decay heat removal by the steam generators or depressurization of the reactor coolant system to increase the water injection rate).

6.54. The emergency core cooling system should be designed with adequate capabilities to prevent or limit the uncovering of the fuel assemblies in the event of primary piping breaks of different sizes and in different locations; different points of injection will usually be necessary.

6.55. The possibility that injection flow rates could bypass the core or could directly flow to the break should be considered when designing the emergency core cooling system.

6.56. The emergency core cooling system should be designed to restore and maintain an adequate coolant inventory in the reactor coolant system in order to recover the fuel cooling function.

6.57. The boron concentration in the emergency core cooling system should be sufficient to achieve core subcriticality in an overcooling design basis accident (e.g. steam line break).

6.58. The injection capacity of the emergency core cooling system should prevent boron crystallization in the core.

6.59. In some designs, the emergency core cooling system includes capabilities to remove core decay heat in the long term when the integrity of the reactor coolant system is not ensured. These capabilities should be considered as part of the safety system.

6.60. The emergency core cooling system should be designed to prevent or limit core uncovering in the event of the total loss of the residual heat removal capacities of the secondary side, taking into account the installed capacity for reactor coolant system bleeding.

*Specific design aspects*

6.61. The pressure retaining equipment for the emergency core cooling system should be designed and manufactured in accordance with proven codes and standards widely used by the nuclear industry (e.g. Refs [24, 25, 27] and similar standards). For each individual component, the requirements to be applied should be selected with due account taken of the two effects resulting from its failure (safety function not fulfilled and a radioactive release occurring).

6.62. The emergency core cooling system should be reliably isolated from the reactor coolant system by two isolation devices in series. In order not to decrease the reliability of the emergency core cooling system, these isolation devices should be designed to open quickly and without external service (e.g. check valves are widely used). The design of the reactor coolant system isolation should enable the leaktightness to be tested periodically. Moreover, the emergency core cooling system should be protected against overpressurization caused by leakages. Adequate means of monitoring (e.g. of pressure and temperature) should be installed to detect and warn about any leakages through the isolation valves.

6.63. The emergency core cooling system equipment should be located outside the containment to the extent possible in order to limit the severity of the environmental conditions for which it should be qualified and also to facilitate maintenance and repair.

6.64. The operation of the emergency core cooling system should limit the risk of causing overpressurization of the reactor coolant system. In particular, cold shutdown states should be considered in which the operation of the emergency core cooling system (spurious or not) could potentially damage the reactor pressure vessel or the residual heat removal system (e.g. owing to brittle fracture).

6.65. Provisions should be implemented for the early detection of leakages in the parts of the emergency core cooling system that are located outside the containment in order to isolate the system before it causes the drainage of the water reserves.

6.66. For accident management, the actuation of the shutdown and isolation of every emergency core cooling system train should be possible from the main control room. However, stopping the operation of the emergency core cooling system from the main control room should not be possible as long as a need for an emergency cooling of the core exists.

6.67. The emergency core cooling system should be qualified to operate with radioactive water loaded with particles, in accordance with the capability of the filtration system.

6.68. The minimum, net positive suction head for the normal operation of the emergency core cooling system pumps should be ensured at any time during design basis accidents, with account taken of limiting phenomena such as vortex, air entrainment and accumulation of debris at the surface of the sump filters. Accounting for the buildup of internal containment pressure should be specifically justified, if national regulatory provisions allow this.

6.69. Mini-flow lines should be implemented to enable periodic tests and to prevent the failure of the emergency core cooling system pumps at low injection flow rates.

**Residual heat removal in hot shutdown modes for design basis accidents**

6.70. The residual heat generated in the core after reactor shutdown is transferred from the reactor coolant system to the steam generators. The heat removal function is ensured by the emergency feedwater system and the steam dump to atmosphere system. The emergency feedwater system comprises emergency feedwater pumps, control valves and isolation valves. The steam dump to atmosphere system comprises a control valve and an isolation valve located at the outlet of the steam generators.

6.71. Paragraphs 6.72–6.81 provide recommendations on meeting Requirement 51 of SSR-2/1 (Rev. 1) [1] in design basis accidents and supplement the recommendations provided in Section 3 relevant for safety systems.

6.72. The emergency feedwater system and the steam dump to atmosphere system should have adequate performance to reliably accomplish residual heat removal and reactor coolant system cooling without exceeding the limits defined for fuel, for the reactor coolant pressure boundary and for structures important to safety in accident conditions without significant fuel degradation.

6.73. The emergency feedwater capacity and autonomy should be sufficient to achieve reactor coolant system conditions for residual heat removal with adequate margins (24 hour autonomy is generally considered as a minimum). Interconnections between emergency feedwater tanks could be considered in order to extend the system capacity and autonomy, provided that the manual operator actions are clearly described in the emergency operating procedures.

6.74. The emergency feedwater system should be designed to supply secondary make-up water to the steam generators in design basis accidents in which the main or auxiliary feedwater system is unavailable.

6.75. The steam dump to atmosphere system should discharge steam from the steam generators in order to remove residual heat and cool down the reactor coolant system when the main condenser is not available or the main steam isolation valves are closed.

*Specific design aspects*

6.76. The pressure retaining equipment for the emergency feedwater system should be designed and manufactured in accordance with proven codes and standards widely used by the nuclear industry (e.g. Refs [24, 25, 27] and similar standards). For each individual component, the requirements to be applied should be selected with due account taken of the consequences resulting from a failure to fulfil its safety function.

6.77. The diversity of the emergency feedwater system pumps should be considered in order to increase the reliability of the system.

6.78. The steam dump valves to atmosphere should be qualified to operate (open and close) for steam, water and a mixture of both (water could be carried by steam if the water level is high in the steam generator).

6.79. Isolation of the emergency feedwater system and of the steam dump valves from the affected steam generator should be performed in the event of a steam line break in order to limit overcooling of the reactor coolant system.

6.80. Isolation of the emergency feedwater system from the affected steam generator should be performed in the event of a steam generator tube rupture in order to prevent the steam generator filling up and to limit the possible release of radioactive water to the environment.

6.81. Isolation of main steam relief valves from the affected steam generator should be performed in the event of a steam generator tube rupture in order to limit the radioactive release to the environment.

**Long term removal of residual heat in design basis accidents**

6.82. The function of the long term residual heat removal system is to transfer residual heat from the reactor coolant system to the intermediate cooling system and achieve a safe shutdown state in accident conditions. This system can be connected after sufficient reactor coolant system cooling. This function is necessary after any design basis accident in which the reactor coolant system water inventory is adequate and controlled.

6.83. The system for the long term removal of residual heat should include several redundant safety systems, each of which includes a pump and a heat exchanger with the intermediate cooling system. The residual heat removal system takes suction water from the reactor coolant system and injects water back into the reactor coolant system after being cooled by the heat exchanger. This system should be considered as the first element of the heat transfer chain to the ultimate heat sink in accident conditions.

6.84. Paragraphs 6.85–6.88 provide recommendations on meeting Requirement 51 of SSR-2/1 (Rev. 1) [1] for the removal of residual heat from the reactor core in design basis accidents.

6.85. The system for the long term removal of residual heat should be designed in accordance with the recommendations provided in paras 3.47–3.50 for safety systems.

6.86. The system for the long term removal of residual heat should be designed to remove core decay heat and to cool down the reactor coolant system to safe shutdown conditions.

6.87. Pressure retaining equipment should be designed and manufactured in accordance with proven codes and standards widely used by the nuclear industry (e.g. Refs [24, 25, 27] and similar standards). For each individual component, the requirements to be applied should be selected with due account taken of the two effects resulting from its failure (safety function not fulfilled and a radioactive release occurring).

6.88. The recommendations provided in paras 3.48–3.52 should also be considered.

**Residual heat removal in hot shutdown modes for design extension conditions without significant fuel degradation**

6.89. Paragraphs 6.90 and 6.91 provide recommendations on meeting Requirement 51 of SSR-2/1 (Rev. 1) [1] for the removal of residual heat from the reactor core in design extension conditions without significant fuel degradation.

6.90. Although the needs for design extension conditions are design dependent, additional design provisions should be considered in order to cope with multiple failures resulting in the loss of the systems and safety systems designed to remove residual heat during reactor coolant system conditions that are not compatible with the residual heat removal operation. Typically, consideration should be given to:

(a)  Extended autonomy of the emergency feedwater system, with on-site refilling capabilities;
(b)  Maintaining the capabilities of the emergency feedwater system and the operation of the steam dump valves to atmosphere in the event of prolonged station blackout;
(c)  Implementation of a secondary side passive heat removal system;
(d)  Removal of decay heat from the core by operating a primary feed and bleed strategy;
(e)  Implementation of a passive system for residual heat removal.

6.91. To facilitate the management of conditions beyond design basis accidents, the emergency feedwater system should include connection lines to supply water into steam generators from external means, for example from fire engines or mobile diesel pumps.

**Fast depressurization of the reactor coolant system in design extension conditions with core melting**

6.92. The reactor coolant system fast depressurization system comprises valves and associated piping, directly connected to the pressure boundary of the reactor coolant system.

6.93. For the practical elimination of the phenomena associated with high pressure melt ejection in severe accidents (direct containment heating), the design should include a fast depressurization of the primary circuit that should be used at the onset of a core melting accident.

*Specific design aspects*

6.94. The reactor coolant system fast depressurization valves should be different and diverse from the safety valves designed for the reactor coolant system overpressure protection.

6.95. The spurious opening of the fast depressurization valves should be reliably prevented.

6.96. Pressure retaining equipment of the reactor coolant system fast depressurization system that is part of the reactor coolant pressure boundary should be designed and manufactured in accordance with proven codes and standards for the design and manufacture of the reactor coolant pressure boundary.

6.97. When the reactor coolant system fast depressurization system is being used in the early phase of a core melting sequence, the temperature and pressure within the reactor coolant system are expected to be very high; therefore, the reactor coolant system fast depressurization system should be designed to operate in such harsh conditions.

6.98. The reactor coolant system fast depressurization system should be designed to withstand SL-2 seismic loads.

6.99. The fast depressurization of the reactor coolant system should be possible in the event of station blackout.

## SYSTEMS FOR CONTROL OF CORE REACTIVITY IN ACCIDENT CONDITIONS

6.100. Paragraphs 6.101–6.106 provide recommendations on meeting the requirements established in paras 6.10 and 6.11 of SSR-2/1 (Rev. 1) [1] with regard to reactor shutdown in accident conditions without significant fuel degradation. These recommendations are for the design of systems that rely on an adequate injection of water with high boric acid concentration. Recommendations for shutdown systems that rely on the drop of solid absorbers are provided in SSG-52 [17].

6.101. The system for control of core reactivity in accident conditions, which is designed as a second and diverse means of shutting down the reactor, should be designed in accordance with the engineering design rules for safety systems.

6.102. The system for control of core reactivity in accident conditions should be independent from the reactor trip system.

6.103. The system for control of core reactivity in accident conditions should have capabilities to shut down the reactor in the event of an anticipated transient without scram without exceeding the fuel limits specified for design extension conditions without significant fuel degradation.

6.104. The boric acid concentration should be sufficient to compensate for the moderator effect variation during reactor coolant system cooling.

**Specific design aspects**

6.105. Pressure retaining equipment should be designed and manufactured in accordance with proven codes and standards widely used by the nuclear industry (e.g. Refs [24, 25, 27] and similar standards). For each individual component, the requirements to be applied should be selected with due account taken of the consequences resulting from a failure to fulfil its safety function.

6.106. Provisions should be considered in normal operation to prevent boron crystallization due to high concentrations in tanks and pipes. Cold conditions, derived from the hazard evaluation for the site, should be considered.

# 7. SPECIFIC CONSIDERATIONS IN DESIGN OF THE ASSOCIATED SYSTEMS FOR BWR TECHNOLOGY

SYSTEMS FOR CONTROL OF COOLANT INVENTORY AND CORE REACTIVITY IN OPERATIONAL STATES

7.1. The control of core reactivity by moving control rods and by controlling the recirculation coolant flow rate is addressed in SSG-52 [17].

**Reactor water cleanup**

7.2. The flow pattern in the reactor pressure vessel should be optimized to prevent the occurrence of areas of stagnant flow (to avoid the accumulation of precipitates) and to limit pockets of cooler water that could result in reactivity excursions or unnecessary thermal stresses.

7.3. The design should also provide for the following:

(a) Limitation of the concentration of contaminants and impurities in the reactor coolant;
(b) Monitoring of the conductivity of the reactor coolant and monitoring for contamination of the sampled fluid;
(c) Control of the water level during shutdown and low power conditions to remove excess water from the reactor coolant system.

7.4. Use of the reactor water cleanup system as the alternate means of residual heat removal during shutdown conditions should be considered.

SYSTEMS FOR HEAT REMOVAL IN OPERATIONAL STATES

7.5. Paragraphs 7.6–7.16 provide recommendations on meeting Requirement 51 of SSR-2/1 (Rev. 1) [1] and supplement the generic recommendations provided in Section 3.

7.6. The design of the plant should include appropriate systems to remove residual heat from the reactor coolant system in different shutdown conditions (e.g. hot shutdown, cold shutdown and during refuelling).

7.7. The heat removal capacity should be designed to cool the reactor coolant system down in a reasonably short period, from hot shutdown conditions (once the reactor is shut down) to conditions adequate for refuelling.

7.8. The minimum heat removal capacity should be designed to remove residual heat even in the event of a single failure postulated in any system necessary for removing residual heat. Moreover, the residual heat removal system should be designed with sufficient capacity such that in the event of a train or division being unavailable during hot shutdown conditions or cold shutdown conditions, the cooldown can continue but at a slower rate.

7.9. The maximal heat removal capacity should be designed taking into account operational criteria (e.g. the time delay to reach refuelling conditions) without exceeding the limits specified in normal conditions for the fuel and the reactor coolant pressure boundary.

7.10. Residual heat removal should be possible in the event of a loss of off-site power.

7.11. Residual heat removal systems should be designed in compliance with all the recommendations provided in Section 3 of this Safety Guide if they are also operated to remove and transfer residual heat after a design basis accident (see para. 2.8).

**Isolation condensers (if included in the design)**

7.12. Isolation condensers should normally be aligned to the reactor coolant system and should be designed in accordance with the same requirements and recommendations that apply to the reactor coolant system.

7.13. The design of the process control valves and the vent valves should be based on the electrical power sources that are assumed to be available at the start of the event. The volume of available water to support the operation of the isolation condensers should be determined by defining the period of time that operation is ensured without refilling the water in the isolation condenser pools. The recommended capability is a minimum of 72 hours but, as a minimum, the capability should meet the requirements of the regulatory body.

7.14. The isolation condenser tubes and tube sheets should be designed for the maximum stresses and most severe fatigue conditions expected to occur in operational states and in design basis accidents.

7.15. The design of the isolation condensers should provide an adequate system for tube leak detection and alarm.

7.16. The design should prevent the accumulation of gases trapped in the steam space of the upper tube sheet and inlet piping of the isolation condenser.


SYSTEMS FOR CORE COOLING AND RESIDUAL HEAT REMOVAL IN ACCIDENT CONDITIONS

**Core cooling in design basis accidents**

7.17. The emergency core cooling system should be designed and implemented to meet Requirement 52 of SSR-2/1 (Rev. 1) [1] in the event of design basis accidents with a loss of primary coolant, as follows:

(a) The emergency core cooling system should be designed to cool the fuel within the limits relevant for fuel and fuel cladding in the event of a loss of coolant accident, taking into account the rules specified for design basis accident analysis.

(b) In case of a small pipe break, the energy removed at the break might not be sufficient for effective cooling of the fuel, and therefore a complementary system or equipment should be operated to achieve the appropriate cooling capacity (e.g. complementary periodic opening of one or more steam relief valves to transfer steam (energy) to the suppression pool), or actuation of the low pressure emergency core cooling system.

(c) The emergency core cooling system should be designed with adequate capabilities to prevent or limit uncovering of the fuel assemblies in the event of primary piping breaks of different sizes.

(d) The design of the overall emergency core cooling system should include both high pressure and low pressure capabilities. The high pressure emergency core cooling system can be used in response to anticipated operational occurrences and small break loss of coolant accidents without the need to depressurize the reactor coolant system.

(e) The emergency core cooling system should be designed to restore and maintain an adequate coolant inventory in the reactor coolant system in order to maintain adequate cooling of the core.

(f) The emergency core cooling system should include capabilities to remove residual heat in the long term, taking into account that the integrity of the reactor coolant system cannot be maintained.

(g)     Owing to the role of the emergency cooling of the core, the systems should be assigned to safety class 1 (see SSG-30 [15]). Individual components should be designed and manufactured in accordance with the engineering requirements given by industry codes (e.g. Refs [24, 25, 27] and similar standards). The requirements to be applied should be selected with due account taken of the two effects resulting from the failure of the component (safety function not fulfilled and a radioactive release occurring).

7.18. The emergency core cooling system should have sufficient capacity to keep the core entirely submerged throughout the most challenging design basis accidents.

7.19. Instrumentation should be provided to control the temperature and water level of the suppression pool during normal operation and in accident conditions.

**Residual heat removal in design basis accidents**

7.20. The design of the plant should include additional systems to remove residual heat from the reactor coolant system in the event of design basis accidents where the systems operated in normal shutdown conditions are not designed to meet the engineering design requirements applicable to safety systems.

**Core cooling in design extension conditions**

7.21. The safety features for design extension conditions are dependent on the reactor technology and the reactor design, and they should be postulated by applying a deterministic approach supported by the outcomes of probabilistic safety assessments.

7.22. The need for additional safety features to ensure the emergency cooling of the core in the event of a loss of coolant accident combined with multiple failures in the emergency core cooling system should be evaluated, and appropriate measures should be implemented as necessary.

7.23. A reliable reactor coolant system depressurization system should be implemented to permit the injection of coolant into the reactor pressure vessel in the event that the high pressure core cooling injection system is unable to maintain an adequate water level in the reactor pressure vessel.

7.24. The capability to cool the core adequately in the event of design extension conditions should be focused on ensuring that such conditions do not result in core melting. As such, the primary focus should be on ensuring that the most probable common cause failure sequences identified for consideration as part of the design extension conditions without significant fuel degradation can be successfully mitigated utilizing on-site equipment.

7.25. The operability of the valves (opening) of turbine driven water supply systems, such as the reactor core isolation cooling system, should be ensured (by using compressed air, DC power or human power) in the event of station blackout (loss of all AC power).

**Reactor coolant system fast depressurization in design extension conditions with core melting**

7.26. A reactor coolant system depressurization system should be implemented to prevent direct containment heating loads caused by the failure of the reactor pressure vessel at high pressure. This function should be accomplished with a different and dedicated set of steam relief valves that should be designed to remain open after the depressurization.

## SYSTEMS FOR CONTROL OF CORE REACTIVITY IN ACCIDENT CONDITIONS

7.27. Recommendations relating to the control rod system are provided in SSG-52 [17].

7.28. The reactor coolant system should have an associated system that is capable of enabling reactor shutdown by injecting a liquid containing a neutron absorbing substance into the reactor pressure vessel; such a system is referred to as the standby liquid control system. This system provides a diverse means of bringing the reactor to a subcritical condition and is used in the event that the control rods cannot be physically inserted into the core.

7.29. The standby liquid control system should have the capability to shut down the core and to maintain subcriticality in the most reactive operational state with an adequate margin.

7.30. The injection rate of the neutron absorbing material should, as a minimum, comply with the requirements of the regulatory body.

7.31. The injection of neutron absorbing material into the reactor pressure vessel should be possible even in the event that off-site power is unavailable.


# 8. SPECIFIC CONSIDERATIONS IN DESIGN OF THE ASSOCIATED SYSTEMS FOR PHWR TECHNOLOGY


8.1. The associated systems are those essential to the safe functioning of the reactor coolant system and connected systems. The associated systems in a PHWR include the following:

(a) The fuel handling system, including the refuelling machines;
(b) The pressure control and inventory control system;
(c) The pump seal cooling system;
(d) The shutdown cooling system;
(e) The emergency core cooling system;
(f) The moderator and its cooling system;
(g) The shield cooling system;
(h) The steam and feedwater system;
(i) The auxiliary feedwater system.


SYSTEMS FOR CONTROL OF CORE REACTIVITY IN OPERATIONAL STATES

8.2. Paragraphs 8.3 and 8.4 provide recommendations on meeting Requirement 45 of SSR-2/1 (Rev. 1) [1] for the systems designed to control reactivity during operational states. The control devices used in the reactor control system include mechanical absorbers, adjusters and light water zone compartments.

8.3. During normal operation, the reactor control system controls the reactor power — and, where applicable, the spatial distribution of the flux — within operating limits for all operational modes. The operational limits can include maximum reactor power, maximum fuel channel (or fuel bundle) power and maximum flux tilt.

8.4. Under anticipated operational occurrences, the reactor control system responds to deviations from normal operation to keep all essential reactor parameters within specified limits and to prevent anticipated operational occurrences from escalating to accident conditions. This response could be reactor power setback or reactor power stepback, which reduce the power to the appropriate levels at different rates.

82

## SYSTEMS FOR HEAT REMOVAL IN OPERATIONAL STATES

8.5. Paragraphs 8.6–8.40 provide recommendations on meeting Requirements 47–53 of SSR-2/1 (Rev. 1) [1] for the systems designed for maintaining the cooling safety function during operational states. In operational states, several systems could be credited for heat removal. These systems comprise the main steam and feedwater system, the shutdown cooling system and the auxiliary feedwater system.

### Steam and feedwater system

8.6. The function of the steam and feedwater system is to transfer the heat produced in the reactor core to the turbine for the generation of power.

8.7. The main steam and feedwater system comprises the main steam lines and the feedwater supply to the steam generators. The main steam lines supply steam from the steam generators in the reactor building to the turbine through the steam balance header, located in the turbine building, at a constant pressure. The feedwater system controls the flow to maintain the necessary steam generator level.

8.8. The main steam and feedwater system should be designed to ensure the stable operation of the reactor at the rated power level during normal operation. The production and dissipation of heat should be balanced at any level of power production.

8.9. The feedwater system should be designed to take hot, pressurized feedwater from the feedwater train in the turbine building and supply it to the secondary side of the steam generators.

8.10. Provision should be made to control the steam pressure and the water inventory in the steam generators during startup.

8.11. The main steam and feedwater system should have sufficient capacity to dissipate heat to the ultimate heat sink during the initial phase of plant cooldown.

8.12. The main steam and feedwater system should have sufficient capacity to dissipate heat to the ultimate heat sink when the main condenser is not available.

8.13. Main steam isolation valves should be provided to isolate the main steam supply to the turbine in the event of steam generator tube leaks after the reactor is shut down, the shutdown cooling system is placed in service and the primary heat transport system is depressurized.

8.14. The safety class (see SSG-30 [15]) of the piping from the steam generators up to and including the main steam isolation valves and the main feedwater isolation valves should be the same as the safety class of the secondary side of the steam generators.

8.15. Redundant heat removal systems should be provided to the extent necessary to enable controlled cooldown of the reactor coolant system when the ultimate heat sink is not available or the main steam line is isolated.

8.16. The main steam and feedwater system should be provided with devices (e.g. safety valves) for the overpressure protection of the secondary side of the steam generators when the main steam and feedwater isolation valves are closed.

8.17. The capacity of the safety valves should be adequate to keep the maximum pressure in the secondary side of the steam generators within the acceptance criteria.

8.18. The system that controls the steam generator pressure should include relief valves. These discharge valves should also provide for overpressure protection to the secondary side of the steam generators in addition to the main steam safety valves.

8.19. As a minimum, the following should be displayed and/or alarmed in the main control room:

(a)     Steam flow rates;
(b)     Steam generator pressure;
(c)     Steam generator level;
(d)     Steam header pressure;
(e)     Feedwater flow rates;
(f)     Feedwater header pressure;
(g)     Feedwater temperature;
(h)     Levels of radioactivity;
(i)     Key chemical parameters.

8.20. Controls should be provided for the main steam isolation valves and bypass valves to allow remote manual operation and automatic operation of these valves.

8.21. Controls should be provided for the valves to dump steam to the atmosphere to allow remote manual and automatic operation.

8.22. The main steam devices (safety and relief discharge valves) should be capable of dissipating heat from the steam generators when the main condenser is not available for heat removal.

8.23. The main steam and feedwater system should provide the capability to automatically and/or manually bypass the turbine and discharge steam directly to the condenser. The capacity of the bypass should be adequate to accommodate the load rejections.

8.24. Provisions should be implemented to ensure that the failure of one steam line will not cause blowdown of the unaffected steam generators.

8.25. Steam lines and feedwater piping should be routed, protected and restrained to prevent multiple accidents in the event of the rupture of a steam line, a feedwater line or any other pipe.

**Shutdown cooling system (residual heat removal system)**

8.26. The shutdown cooling system comprises pumps and heat exchangers connected between the inlet and outlet headers of each primary heat transport system loop. The system is normally full of heavy water and is normally isolated from the primary heat transport system.

8.27. The function of the shutdown cooling system is to provide fuel cooling for an indefinite period of time after reactor shutdown. It is also designed to provide the cooling function when the primary heat transport system to the reactor headers is drained to permit maintenance of the steam generators and the primary heat transport system pump internals.

8.28. The shutdown cooling system should preferably be located inside the containment.

8.29. The shutdown cooling system should have the capability to control the heavy water level in the primary heat transport system headers in the drained state.

8.30. The shutdown cooling system should have the capability to cool down the primary heat transport system in the event that heat removal via the steam generators is suddenly not possible.

8.31. The shutdown cooling system should be designed to also remove residual heat when the reactor is shut down following an accident, by functioning as an alternative heat removal to the steam generators.

8.32. The shutdown heat exchangers should be designed to withstand extreme temperature shocks.

8.33. The shutdown cooling system should allow the lowering, raising and controlling of the level of coolant in the reactor coolant system to allow maintenance of the heat transport pumps and the steam generators. The shutdown cooling system should have the capability to be used for draining the primary heat transport system when this heat transport system is cold and depressurized.

8.34. The shutdown cooling system should have sufficient flow adjustment capability.

8.35. To ensure the integrity and reliability of the shutdown cooling system, provisions should be implemented in its design and layout to enable the inspection of major components to be carried out during outages.

8.36. The heat transfer should be ensured in operational states and related postulated accident conditions. Ranges of relevant parameters (e.g. temperature and pressure of the primary coolant) should be specified for each plant state.

**Auxiliary feedwater system**

8.37. The feedwater system could comprise the following systems:

(a)     A main feedwater system;
(b)     An auxiliary feedwater system;
(c)     An emergency heat removal system.

8.38. An auxiliary feedwater system or equivalent should be designed to maintain the heat removal capability of the plant in the event that the main feedwater system becomes unavailable. The capability of the auxiliary feedwater system for heat removal might also be used to reduce the pressure in the reactor coolant system when necessary.

8.39. An auxiliary feedwater system or equivalent should be designed to maintain the plant in a hot standby condition for an extended period. The auxiliary feedwater system should provide sufficient capacity to fulfil this function efficiently. Where a connection to the reserve feedwater or to the deaerator is not possible, an alternate means to supply the auxiliary feedwater to the steam generators should be provided.

8.40. The design of the auxiliary feedwater system should include connection lines to supply water to the steam generators from the reserve water tank (also called the containment water tank or the dousing reservoir) and from fire engines or mobile diesel pumps. Means for recording the amount of water supplied to the steam generators should be provided.

## SYSTEMS FOR CONTROL OF CORE REACTIVITY IN ACCIDENT CONDITIONS

8.41. Paragraphs 8.42–8.49 provide recommendations for systems designed to fulfil the safety function of reactivity control during accident conditions, including design basis accidents and design extension conditions without significant fuel degradation.

### Reactor shutdown systems

8.42. A PHWR reactor is equipped with two physically independent shutdown systems. These shutdown systems are designed to be both functionally different and geometrically separate. The functional difference is achieved by the use of shut-off rods for the first shutdown system and the injection of a liquid neutron absorber (poison) for the second shutdown system. Both shutdown systems are designed to be independently capable of quickly rendering the reactor subcritical from all anticipated operational occurrences and design basis accidents by an adequate margin.

8.43. The second reactor shutdown system provides a fast injection of a liquid neutron absorber into the bulk moderator through a number of horizontally distributed nozzles. The second reactor shutdown system employs independent multiple reactor trip logics that identify the parameters for shutdown and open fast acting valves to inject the poison into the moderator.

8.44. The first and second reactor shutdown systems should be fast acting, fully capable, diverse and functionally independent of each other. If these systems are part of the reactor design, they should also be passive.

8.45. The second reactor shutdown system should be designed to avoid issues relating to chemistry (e.g. avoiding precipitation).

8.46. The second reactor shutdown system should be designed to meet the acceptance criteria for the reactor trip parameter effectiveness for all anticipated operational occurrences and accident conditions without significant fuel degradation.

8.47. In the 'poised state' (i.e. being capable of adding sufficient negative reactivity to shut down the reactor), the second reactor shutdown system should be designed:

(a) To hold outside the reactor core a sufficient amount of poison with an appropriate concentration, chemical composition and absorbing properties, ready to be injected into the moderator to shut down the reactor;

(b) To provide means of verifying the amount of poison and its concentration, correct chemical composition and absorbing properties;

(c) To provide means of injecting, as efficiently and effectively as possible, the poison into the moderator following a reactor trip signal in the second reactor shutdown system;

(d) To provide means of backflushing the injection lines in which the poison concentration is excessive as a consequence of poison migration.

8.48. In the 'tripped state' (i.e. when the poison has been injected into the moderator to shut down the reactor and to maintain it subcritical), the second reactor shutdown system should be designed:

(a) To be capable of maintaining the reactor subcritical following a second reactor shutdown system reactor trip;

(b) To allow the injection to be stopped if the second reactor shutdown system trip logic clears an unsealed in-reactor trip state.

8.49. The reliability criteria should cover sensing the need for shutdown, initiation of shutdown and insertion of negative reactivity. All elements necessary to complete the shutdown function should be included.

## SYSTEMS FOR CORE COOLING AND RESIDUAL HEAT REMOVAL IN ACCIDENT CONDITIONS

8.50. Paragraphs 8.51–8.117 provide recommendations on meeting Requirements 47–53 of SSR-2/1 (Rev. 1) [1] for the systems designed to fulfil the cooling safety function in accident conditions, including design basis accidents and design extension conditions without significant fuel degradation. Systems designed for maintaining the cooling safety function for design basis accidents include the emergency core cooling system and the enhanced emergency heat removal system.

8.51. Systems in the reactor coolant system and associated systems that are provided to mitigate the consequences of design basis accidents should be considered safety systems and should be designed in accordance with the engineering rules established for these systems.

8.52. Systems that are provided to mitigate the consequences of design extension conditions should be considered safety features for design extension conditions and should be designed in accordance with the engineering rules established for these features. These systems could be designed with appropriate redundancy within the systems to achieve the necessary reliability.

**Emergency core cooling system**

8.53. The emergency core cooling system includes a combination of active and passive injection means (pumps, piping and valves) with different delivery pressures, depending on the designs, and also passive injection tanks (accumulators). The system might also include heat exchangers.

8.54. The emergency core cooling system supplies cooling water (light water) to the reactor coolant system following a loss of coolant accident in which the inventory of heavy water is lost. It should be designed to remove residual heat from the reactor.

8.55. The emergency core cooling system should be designed to cool the core adequately in the event of a double ended guillotine break of a header.

8.56. The injection capacity of the emergency core cooling system should ensure core reflooding in the event of a design basis loss of coolant accident, in accordance with the applicable acceptance criteria.

8.57. The emergency core cooling system should be capable of maintaining the core in a coolable geometry and of removing residual heat.

8.58. If the injection pressure of the emergency core cooling system is lower than the opening pressure of the steam generator relief devices, it will limit releases from the active steam generators in the event of design basis steam generator tube rupture. In any case, the injection pressure should be lower than the opening pressure of the steam generator safety valves in order to limit the risk that they open and then fail to close.

8.59. The injection pressure of the emergency core cooling system should limit the risk of causing reactor coolant system overpressure.

8.60. The injection of a large volume of cold water might cause pressurized thermal shock to the reactor coolant pressure boundary or distortion of reactor internals, especially in cold shutdown states. It should be demonstrated that thermal shock has been addressed in the design, in terms of calculating the transient fluid conditions at key locations, the resulting metal temperature and the corresponding stresses.

8.61. The emergency core cooling system can also provide residual heat transfer to the heat exchanger by cooling the sumps in the event of design basis accidents, especially large break loss of coolant accidents. The heat exchanger capacity should be sufficient to limit sump heat-up to within a temperature range compatible with the qualification conditions inside the reactor building and with the qualification of the emergency core cooling system pumps.

8.62. Because the emergency core cooling system is connected to the reactor coolant system, it should be equipped with isolation devices as are necessary for the reactor coolant pressure boundary. These devices (e.g. isolation valves) should be closed in normal operation and should open quickly in case injection is needed. It should be possible to reclose them if the injection is stopped in the long term after an accident, especially if there is a suspected leak in a system train.

8.63. As the emergency core cooling system is partly located outside the containment, it is required to be equipped with containment isolation devices, in accordance with Requirement 56 of SSR-2/1 (Rev. 1) [1]. To prevent the drainage of the sump, it should be possible to close the suction valves from the sumps at any time if a leak is detected in the part of the system located outside the containment. Such isolation should be performed with a high level of confidence, as its failure would lead to a severe accident with total depletion of the water reserves.

8.64. All the components of the emergency core cooling system that belong to the reactor coolant pressure boundary (e.g. injection nozzles) should be designed taking into consideration the same quality requirements and the same loads as the reactor coolant system pipes.

8.65. Unplanned drainage of the emergency core cooling system water reserve should be prevented, specifically in case of external hazards. The containment penetrations of the suction pumps and the isolation valves of the emergency core cooling system should be adequately protected.

8.66. The emergency core cooling system pumps should be qualified to operate with radioactive water loaded with particles, in accordance with the capability of the filtration system. The qualification specification should consider the levels of radioactivity and debris release assumed for design basis accidents and design extension conditions without significant fuel degradation (loss of coolant accidents and secondary breaks).

8.67. Monitoring for possible leakages in the parts of the emergency core cooling system that are located outside the containment should be implemented in order to be able to isolate the system before the leak causes the drainage of the water reserves, and before it causes environmental conditions in the building that would preclude the operation of the isolation valves.

8.68. The emergency core cooling system isolation devices that are located outside the containment should be qualified to remain operable even in the event of a possible leak from the system.

8.69. Natural circulation flows, where credited, should be capable of providing sufficient flow and should not be impaired by effects such as the accumulation of non-condensable gas or adverse temperature distributions.

8.70. In the event of a loss of coolant accident, the local effect of the break (e.g. pipe whip or jet impingement, including potential blast waves) should be limited so that no more than one emergency core cooling system train is made unavailable.

8.71. The recirculation pumps of the emergency core cooling system should be located outside the containment in order to limit the severity of the environmental conditions for which they should be qualified and also to facilitate maintenance and repair.

8.72. The emergency core cooling system provides an extension of the containment (a third barrier) when circulating water outside the containment in the event of design basis accidents. This water could be highly radioactive, for example in the event of damage to the fuel; therefore, the structural design recommendations for components should preclude radioactive releases with a high degree of confidence (see also the recommendations provided in SSG-53 [22]).

8.73. The pumps in the emergency core cooling system might require motor and room cooling for proper operation. These support functions should be performed with a reliability level commensurate with their importance. If their failure might lead to pump failure in too short a time to set up alternative cooling, then the design recommendations for these systems should be consistent with those for the emergency core cooling system.

8.74. The cooling of the emergency core cooling system pumps by diversified means could be considered if their injection function is necessary in design extension conditions where a common cause failure of the means of cooling is assumed.

**Enhanced emergency heat removal system**

8.75. The function of the enhanced emergency heat removal system is to feed water to the steam generators in order to maintain heat removal capability.

8.76. The design of the emergency heat removal system should ensure that there is adequate long term heat transfer available for residual heat removal following a loss of the normal heat removal systems (main and auxiliary feedwater).

8.77. The emergency heat removal system should have independent passive (backup emergency feedwater) and active (emergency feedwater) trains. Each of the active and passive trains of the emergency heat removal system should have the capability to maintain a sufficient water inventory in the secondary side of the steam generators.

8.78. The active train of the emergency heat removal system and its supporting structures, systems and components should be designed to operate under the postulated initiating events considered for design basis accidents and which result in the loss of the normal heat removal systems.

8.79. The active train of the emergency heat removal system that is provided to mitigate the consequences of design basis accidents should meet the design requirements for safety systems.

8.80. The passive train of the emergency heat removal system and its supporting structures, systems and components should be designed to operate under design extension conditions without significant fuel degradation.

8.81. The active (emergency feedwater) train of the emergency heat removal system could have pumps taking suction from a source of on-site fresh water that is in a separate location from the main service water system intake for the plant. This active train of the emergency heat removal system should have an emergency power supply with an automated startup, and connection lines to supply water to the secondary side of the steam generators.

8.82. The passive (backup emergency feedwater) train of the emergency heat removal system should consist of the reserve water tank (also called the containment water tank or the dousing reservoir) and connection lines (including valves and piping) to supply water to the secondary side of the steam generators.

8.83. The reserve water tank should be designed as a gravity driven, passive, light water make-up system, such that no external power is needed to transfer its inventory to the various potential destinations once the isolation valves are opened.

8.84. The reserve water tank should be located at a high elevation in the reactor building.

8.85. The reserve water tank should have sufficient capacity to provide an emergency source of water by gravity to the steam generators (backup emergency feedwater), the containment cooling spray, the moderator system, the shield cooling system and the primary heat transport system, as necessary.

8.86. The active and passive trains of the emergency heat removal system need to be functional during and after a seismic event and therefore should be designed to meet the seismic requirements.

8.87. The design should demonstrate that emergency heat removal capability is provided for all operational states and all accident conditions.

8.88. Equipment should be appropriately designed to function in the class of accidents for which it is credited, for all means of emergency heat removal.

8.89. The design should provide provisions to allow in-service inspection of components and equipment and to allow operational functional testing of systems and components.

8.90. The emergency heat removal system should be capable of removing the heat loads from structures, systems and components in design basis accidents and design extension conditions without significant fuel degradation.

8.91. The application of the single failure criterion for all safety features for design extension conditions is not explicitly required by SSR-2/1 (Rev. 1) [1].

8.92. The appropriate emergency power supply (AC or DC) should be provided, as necessary, to components that are needed for the actuation or operation of safety features for design extension conditions.

8.93. The safety features for design extension conditions should be qualified such that they will function under the most severe environmental conditions (including seismic conditions) in which they would be expected to operate.

8.94. Manual actuation of the safety features for design extension conditions should be possible from the main control room and if appropriate from the supplementary control room.

8.95. Process information and control capability should be provided in the main control room and in the supplementary control room to enable the passive and active trains of the emergency heat removal system to be operated and to achieve adequate reactor residual heat removal on a long term basis.

8.96. To ensure the integrity and reliability of the emergency heat removal system, provisions in the design and layout should be implemented to enable the inspection of major components during outages.

8.97. The capability of natural circulation systems should be demonstrated over the full range of applicable operating conditions.

8.98. The need for automatic actuation of safety features for design extension conditions without significant fuel degradation should be evaluated on a case by case basis in the safety analysis.

8.99. In cases where the active train of the emergency heat removal system is credited for design basis accidents, an analysis should be performed to demonstrate that the acceptance criteria are met. The analysis should be performed with adequate conservatism to demonstrate that the margins provided in the design are appropriate to accommodate uncertainties and to prevent cliff edge effects.

8.100.   In cases where the passive train of the emergency heat removal system is credited for design extension conditions without significant fuel degradation, an analysis should be performed to demonstrate that the acceptance criteria are met. The best estimate analysis methodology is acceptable.

**Heat transfer in design extension conditions**

8.101.   Provisions for heat transfer in design extension conditions should be provided by complementary safety features that have the capability to transfer residual heat from the core to an ultimate heat sink. These features: should be independent, to the extent practicable, of the design features used in more frequent accidents; should be capable of performing in the environmental conditions pertaining to design extension conditions; and should have a reliability commensurate with the safety function that they are expected to fulfil.

8.102.   The design principles for safety features for design extension conditions do not necessarily need to incorporate the same degree of conservatism as those applied to the design for operational states and for design basis accidents. However, there should be reasonable assurance that safety features for design extension conditions will function as designed when called upon.

8.103.   The design rules for such complementary safety features for design extension conditions should be clearly described and should be derived from operating experience, the latest results from safety related research and development and up to date design practices.

**Moderator system in design extension conditions without significant fuel degradation**

8.104.   The moderator system of a PHWR reactor is a low pressure and low temperature system, and is independent of the primary heat transport system. It comprises pumps and heat exchangers that circulate the heavy water moderator through the calandria and remove the heat generated during reactor operation. For normal operation and design basis accidents, the heavy water acts as both the moderator and reflector for the neutron flux in the reactor core.

8.105.   The moderator system should have its own cooling system to remove heat transferred from the reactor structure and the heat generated by radioactive decay in the moderator system.

8.106.   The moderator system fulfils a safety function that is unique to PHWRs. The moderator system should be designed to act as a means of emergency heat removal for design extension conditions without significant fuel degradation under the postulated accident condition of a large loss of coolant accident coincident with the loss of the emergency core cooling system.

8.107.   The design of the moderator system should consider all system configurations when credited as an emergency heat removal system for design extension conditions without significant fuel degradation. Each configuration should have an adequate load capacity to independently transfer heat to the ultimate heat sink and to prevent the failure of the calandria tubes.

8.108.   The heat load capability of the moderator system configuration for design extension conditions without significant fuel degradation should be demonstrated by means of tests and analyses.

8.109.   The moderator system should be designed such that forced convection flows and natural convection flows are in the same direction.

8.110.   The moderator system components should be designed and built to higher standards than are otherwise necessary in order to minimize the possibility of heavy water loss and to maximize reliability.

8.111.   The moderator pumps should be designed to retain their pressure integrity during and following a design basis earthquake.

8.112.   The moderator system should be designed for overpressure protection from the pressure transients arising in the calandria from the burst of a pressure tube or a calandria tube.

8.113.   The calandria vessel should be equipped with overpressure protection devices such as rupture discs or equivalent devices.

8.114.   The relief capacity should be sufficient to avoid overpressurization limits for the structures, systems and components credited for design extension conditions without significant fuel degradation. Limits prescribed by proven codes and standards applicable to nuclear pressure vessels should be used.

**Provisions for fast depressurization of the primary heat transport system (crash cooldown)**

8.115.    PHWRs should be equipped for a fast depressurization of the primary circuit by the crash cooldown of the steam generator secondary side (or equivalent) using the steam relief valves.

8.116.    The design should demonstrate that during crash cooldown:

(a)    The inventory of the secondary side of the steam generators will be sufficiently maintained to cool and depressurize the reactor coolant system;
(b)    The water inventory in the reactor coolant system will be maintained;
(c)    The heat transfer mechanism (e.g. thermo-syphoning or intermittent buoyancy induced flow) in the reactor coolant system will not be disrupted.

8.117.    A crash cooldown or a reactor coolant system depressurization should not result in any reactivity or structural concerns.


# REFERENCES


[1]    INTERNATIONAL ATOMIC ENERGY AGENCY, Safety of Nuclear Power Plants: Design, IAEA Safety Standards Series No. SSR-2/1 (Rev. 1), IAEA, Vienna (2016).

[2]    INTERNATIONAL ATOMIC ENERGY AGENCY, IAEA Safety Glossary: Terminology Used in Nuclear Safety and Radiation Protection, 2018 Edition, IAEA, Vienna (2019).

[3]    INTERNATIONAL ATOMIC ENERGY AGENCY, Leadership and Management for Safety, IAEA Safety Standards Series No. GSR Part 2, IAEA, Vienna (2016).

[4]    INTERNATIONAL ATOMIC ENERGY AGENCY, Application of the Management System for Facilities and Activities, IAEA Safety Standards Series No. GS-G-3.1, IAEA, Vienna (2006).

[5]    INTERNATIONAL ATOMIC ENERGY AGENCY, The Management System for Nuclear Installations, IAEA Safety Standards Series No. GS-G-3.5, IAEA, Vienna (2009).

[6]    INTERNATIONAL ATOMIC ENERGY AGENCY, Nuclear Security Recommendations on Physical Protection of Nuclear Material and Nuclear Facilities (INFCIRC/225/ Revision 5), IAEA Nuclear Security Series No. 13, IAEA, Vienna (2011).

[7]    INTERNATIONAL ATOMIC ENERGY AGENCY, Considerations on the Application of the IAEA Safety Requirements for the Design of Nuclear Power Plants, IAEA-TECDOC-1791, IAEA, Vienna (2016).

[8]    INTERNATIONAL ATOMIC ENERGY AGENCY, Protection against Internal Fires and Explosions in the Design of Nuclear Power Plants, IAEA Safety Standards Series No. NS-G-1.7, IAEA, Vienna (2004). (A revision of this publication is in preparation.)

[9]    INTERNATIONAL ATOMIC ENERGY AGENCY, Protection against Internal Hazards other than Fires and Explosions in the Design of Nuclear Power Plants, IAEA Safety Standards Series No. NS-G-1.11, IAEA, Vienna (2004). (A revision of this publication is in preparation.)

[10]   INTERNATIONAL ATOMIC ENERGY AGENCY, External Events Excluding Earthquakes in the Design of Nuclear Power Plants, IAEA Safety Standards Series No. NS-G-1.5, IAEA, Vienna (2003). (A revision of this publication is in preparation.)

[11]   INTERNATIONAL ATOMIC ENERGY AGENCY, Evaluation of Seismic Safety for Existing Nuclear Installations, IAEA Safety Standards Series No. NS-G-2.13, IAEA, Vienna (2009).

[12]   INTERNATIONAL ATOMIC ENERGY AGENCY, Seismic Design and Qualification for Nuclear Power Plants, IAEA Safety Standards Series No. NS-G-1.6, IAEA, Vienna (2003). (A revision of this publication is in preparation.)

[13]   INTERNATIONAL ATOMIC ENERGY AGENCY, Seismic Hazards in Site Evaluation for Nuclear Installations, IAEA Safety Standards Series No. SSG-9, IAEA, Vienna (2010). (A revision of this publication is in preparation.)

[14]   INTERNATIONAL ATOMIC ENERGY AGENCY, Design of Electrical Power Systems for Nuclear Power Plants, IAEA Safety Standards Series No. SSG-34, IAEA, Vienna (2016).

[15]   INTERNATIONAL ATOMIC ENERGY AGENCY, Safety Classification of Structures, Systems and Components in Nuclear Power Plants, IAEA Safety Standards Series No. SSG-30, IAEA, Vienna (2014).

[16]   INTERNATIONAL ATOMIC ENERGY AGENCY, Operational Limits and Conditions and Operating Procedures for Nuclear Power Plants, IAEA Safety Standards Series No. NS-G-2.2, IAEA, Vienna (2000). (A revision of this publication is in preparation.)

[17]   INTERNATIONAL ATOMIC ENERGY AGENCY, Design of the Reactor Core for Nuclear Power Plants, IAEA Safety Standards Series No. SSG-52, IAEA, Vienna (2019).

[18]   INTERNATIONAL ATOMIC ENERGY AGENCY, Design of Fuel Handling and Storage Systems for Nuclear Power Plants, IAEA Safety Standards Series No. SSG-63, IAEA, Vienna (in preparation).

[19]   INTERNATIONAL ATOMIC ENERGY AGENCY, Ageing Management and Development of a Programme for Long Term Operation of Nuclear Power Plants, IAEA Safety Standards Series No. SSG-48, IAEA, Vienna (2018).

[20]   INTERNATIONAL ATOMIC ENERGY AGENCY, Maintenance, Surveillance and In-service Inspection in Nuclear Power Plants, IAEA Safety Standards Series No. NS-G-2.6, IAEA, Vienna (2002). (A revision of this publication is in preparation.)

[21]   INTERNATIONAL ATOMIC ENERGY AGENCY, Radiation Protection Aspects of Design for Nuclear Power Plants, IAEA Safety Standards Series No. NS-G-1.13, IAEA, Vienna (2005).

[22]  INTERNATIONAL ATOMIC ENERGY AGENCY, Design of the Reactor Containment and Associated Systems for Nuclear Power Plants, IAEA Safety Standards Series No. SSG-53, IAEA, Vienna (2019).

[23]  INTERNATIONAL ATOMIC ENERGY AGENCY, Design of Instrumentation and Control Systems for Nuclear Power Plants, IAEA Safety Standards Series No. SSG-39, IAEA, Vienna (2016).

[24]  AMERICAN SOCIETY OF MECHANICAL ENGINEERS, ASME Boiler and Pressure Vessel Code, Section III, Division 1, Rules for Construction of Pressure Vessels, ASME, New York (2013).

[25]  AFCEN, RCC-M: Design and Construction Rules for Mechanical Components of PWR Nuclear Islands, AFCEN, Paris (2019).

[26]  INTERNATIONAL ATOMIC ENERGY AGENCY, Application of the Safety Classification of Structures, Systems and Components in Nuclear Power Plants, IAEA-TECDOC-1787, IAEA, Vienna (2016).

[27]  JAPAN SOCIETY OF MECHANICAL ENGINEERS, JSME Codes for Nuclear Power Generation Facilities: Rules on Design and Construction for Nuclear Power Plants, JSME, Tokyo (2015) (in Japanese).

# CONTRIBUTORS TO DRAFTING AND REVIEW

| | |
|---|---|
| Baik, S.J. | KEPCO Engineering and Construction Company, Republic of Korea |
| Beard, J. | GE Hitachi Nuclear Energy, United States of America |
| Courtin, E. | AREVA, France |
| Fil, N. | Consultant, Russian Federation |
| Gasparini, M. | Consultant, Italy |
| Jackson, C. | Nuclear Regulatory Commission, United States of America |
| Mesmous, N. | Canadian Nuclear Safety Commission, Canada |
| Myeong-Yong Ohn | Canadian Nuclear Safety Commission, Canada |
| Nakajima, T. | Nuclear Regulation Authority, Japan |
| Poulat, B. | International Atomic Energy Agency |
| Taniguchi, A. | Tokyo Electric Power Company Holdings, Japan |
| Toth, C. | International Atomic Energy Agency |
| Yamazaki, H. | Toshiba Corporation, Japan |
| Yllera, J. | International Atomic Energy Agency |
| Yoshikawa, K. | Hitachi-GE Nuclear Energy, Japan |

International Atomic Energy Agency



# ORDERING LOCALLY

IAEA priced publications may be purchased from the sources listed below or from major local booksellers.

Orders for unpriced publications should be made directly to the IAEA. The contact details are given at the end of this list.

## NORTH AMERICA

**Bernan / Rowman & Littlefield**
15250 NBN Way, Blue Ridge Summit, PA 17214, USA
Telephone: +1 800 462 6420 • Fax: +1 800 338 4550
Email: orders@rowman.com • Web site: www.rowman.com/bernan

**Renouf Publishing Co. Ltd**
22-1010 Polytek Street, Ottawa, ON K1J 9J1, CANADA
Telephone: +1 613 745 2665 • Fax: +1 613 745 7660
Email: orders@renoufbooks.com • Web site: www.renoufbooks.com

## REST OF WORLD

Please contact your preferred local supplier, or our lead distributor:

**Eurospan Group**
Gray's Inn House
127 Clerkenwell Road
London EC1R 5DB
United Kingdom

**Trade orders and enquiries:**
Telephone: +44 (0)176 760 4972 • Fax: +44 (0)176 760 1640
Email: eurospan@turpin-distribution.com

**Individual orders:**
www.eurospanbookstore.com/iaea

**For further information:**
Telephone: +44 (0)207 240 0856 • Fax: +44 (0)207 379 0609
Email: info@eurospangroup.com • Web site: www.eurospangroup.com

**Orders for both priced and unpriced publications may be addressed directly to:**
Marketing and Sales Unit
International Atomic Energy Agency
Vienna International Centre, PO Box 100, 1400 Vienna, Austria
Telephone: +43 1 2600 22529 or 22530 • Fax: +43 1 26007 22529
Email: sales.publications@iaea.org • Web site: www.iaea.org/publications

Printed and bound by CPI Group (UK) Ltd, Croydon, CR0 4YY

06/07/2026

02160597-0004